Bending The Ruler

Time Travel, The Speed of Light, Gravity, and The Big Bang

BENDING THE RULER

R Lindemann

Aleph Publications
Wisconsin, USA

Bending The Ruler
Time Travel, The Speed of Light, Gravity, and The Big Bang
Copyright 2011 - R Lindemann ©
All Rights Reserved. Published 2023

Aleph Publications
Manitowoc, WI

Paperback Edition
ISBN-13: 978-0-9893318-8-3

33 32 31 30 29 28 27 26 25 24   2 3 4 5 6

Dedication

This book is dedicated to all of the great minds who have departed, and to those who are still with us, who have broken free of conventional thinking in effort to discover the truth about our Universe and all that is contained within. Your courage has been an inspiration to many!

This book is also dedicated to the fresh young upcoming minds, and to those yet to be born, who will seek to know the truth. As you grow, respect the minds of our past thinkers, but break through what they could not yet see. Discover for yourself, and for the rest of humanity, the wonders that past great thinkers all sought to know. When you find those and other treasures, share those treasures in selfless honesty with the world!

Disclaimer

All information, views, thoughts, and opinions expressed herein are those of the author(s) and are being presented only for your consideration and should not be interpreted as advice to take any action. Any action you take with regard to implementing or not implementing the information, views, thoughts, and opinions contained within this published work is your own responsibility. Under no circumstances are distributor(s) and/or publisher(s) and/or author(s) of this work liable for any of your actions.

Anyone, especially those who have been victim of misdirected explanation and understanding, may be best served seeking wise counsel before deciding to implement any information, views, thoughts, opinions, or anything else that is offered for your consideration in this work. All information, views, thoughts, and opinions in this work are not advice, directive, recommendation, counsel, or any other indication for anyone to take any action. All information, views, thoughts, and opinions offered herein are offered only as suggestions for your personal consideration, which is done of your own free will. Your life is your own responsibility; use it wisely.

Any use of trade names or mention of commercial sources is for informational purposes only and does not imply endorsement or affiliation.

Contents

Figures

Acknowledgments

My curiosity about space, space-travel, and physics has always been there from early on. While I didn't understand what "physics" was while in my early youth, contemplating ideas such as space, time, gravity, and infinity was common for me. Yes, it's odd for a six- or seven-year-old to wonder about these things, but I would always listen intently when hearing things about Copernicus, Galileo, Newton, and even Einstein. Great thinkers and discoverers like Copernicus, Newton, and Galileo are to be credited for much of what we understand about science today; so, I would like thank them, posthumously, for their wonderful contributions to our understanding of the heavens. I cannot say whether or not my curiosity would have left me without their contributions, but their thoughts certainly made it easier for all of us today.

And for more current events, I also want to offer my most sincere and deepest thanks to everyone who helped in preparing, editing, printing, and distributing this book. The careful debates and deep discussions with those who offered editorial thoughts kept me in check to make sure this book assesses past theories in a fair manner. Special thanks to Halie Hackbarth and to Lisa A Miller for your helpful reviews, edits, and discussions. And to everyone involved, thank you for your patience and dedication, you all have been wonderfully supportive!

Introduction

As a child, science and physics were intriguing and were of natural interest to me. Curiosity about how things worked was of no small matter to my parents, as I dismantled anything I could get my hands on in order to understand how it worked.

In the early adult years, I had a naïve grasp of our world in that the view of simplicity in all that was around us was shared by everyone. But as realized later on in life, this was not so. The hesitation from getting too deeply into scientific discussion in the earlier years was because most of my work was in mechanical, computer, and chemical research and development. Yet the many unanswered questions about science, physics, and astrophysics kept persisting.

Whenever opportunity arose, I would read up on various areas of scientific study and have always found my deepest curiosity to lie within the physics realm, especially in relation to time, time-travel, the speed of light, and gravity. These exciting areas of study are thought by many to be mastered. But as you will see, what we believe today to be the absolute "truth" about science is not as absolute as we think.

This book has been written because the world, especially science-minded youth, need to understand that we have only just begun to understand the physical world around us. This book is written in simple language so that everyone can read it without feeling as if it is beyond their scope of knowledge and understanding.

In order to expose the errors that stop us from advancing towards the actual truth, we must bring to light, the many inconsistencies seen and practiced in the realm of science. Within these pages there are perspectives from both sides of an issue, not totally committing to one or the other. Yes, I have my own personal views with regard to both science and religion, but am always excited and ready to hear contrary ideas. Contrary ideas are always welcome because contrary ideas are the best ideas to use in order to test any views we hold in *open* conclusion.

This book is not about the specifics of science, it's about opening our minds so that we can achieve the next level of science. For instance, if we are unwilling to say that Einstein may have been wrong or maybe that we understand him incorrectly, then we are trapped in whatever wrongness he may have practiced or the wrongness we may have interpreted from his conclusions.

In order to search out the absolute truth, science needs fresh new bright young minds to step up and see through society's preconceptions of what we "believe" is correct.

Challenge yourself to set aside all of your preconceptions and leave your mark on the world of science by teaching the rest of humanity to experience your perspective through our own eyes—find truth. This will allow humanity to see the beauty within each of your discoveries!

Chapter 1

Understanding Our Universe

Since the earliest of recorded history, we humans have left evidence of our desire to understand our origins and what we see around us. It seems that each day we are ever closer to understanding our origins, but are we? There are answers to the questions of "The Beginning," but are these answers of a religious nature, a scientific nature, or maybe both?

For thousands of years, philosophers, religious leaders, and scientific-minded people have been seeking to understand our origins and how the "Heavens" came to be. Some of their ideas have been proven wrong, some are still unknown, and other theories are believed to be correct; however, an infinite amount of ideas have yet to be conceived! Regardless of whether the minds of the past were right or wrong, the underlying goal of almost everyone has been the same in this regard—we all want to know where we are from, how it all works, and how everything came to be.

Each subsequent generation of thinkers builds upon the successes and failures of the previous generations of thinkers. It is my goal to bring many new people deep into the quest to understand how everything that is in existence came to be. I believe that humanity can achieve great understanding from our past observations and scientific experiences—even if some of them are wrong. But here in this book you are offered views, thoughts, and opinions for consideration that might contradict your current understanding of things. All that anyone can do is to offer views, thoughts, and opinions for consideration, but what *you* do with it and the direction it takes you is your own responsibility and is a decision that is in your hands alone. If you choose to follow a wrong direction then you will be wrong and you will have difficulty overcoming the blindness that your error has set upon you. We have not yet even begun to scratch the surface of understanding how everything came to be. The more that we observe, then the more we can see the grand infinite beauty of it all. Open your eyes and see all that there is to see!

Practical Application Matters to You

Science is faced with the single most intrusive stumbling-block ever seen by mankind. We will not be able to understand our origins if we cannot protect ourselves from the blindness caused by the religious- and scientific-style assumptions that trip us up.

Following is a brief example to help you to understand this troubling blindness more clearly: Some years prior to writing this book I spoke to an engineer at a manufacturer of electric motor servo drives. He knew the math and didn't seem to understand or accept practical application. We were calculating the value of the forces generated in the inertial state of a load-bearing servo motor.

The engineer entered our data into his computer program which showed that the motor and drives could not bear the presented load. Our load mass was very light and very near to the axis of the motor's shaft. He insisted that the motor could not handle our specified load at the acceleration rates that we required, but to our staff this was clearly wrong. After some discussion, we completely removed the load from the equation and found that the computer program used to calculate the energy needed for a zero-load, showed that the motor alone could not accelerate at the desired speed. While this is a possibility it was not necessarily true because the computer program quite obviously did not have zero-load as a consideration in its calculations.

It is important to convey to you that from *practical application* experience, it was understood that the motor alone could accelerate at the required rate, and also that the motor could handle the load's mass. The engineer at the motor company was not a stupid person, but he had an "educated" blindness that was due, in part, to his dependency on computers and the rigid math reasoning from his college education. These blindnesses did not allow him the ability to fully understand the particular application or believe that it was possible. Nothing seemed to get him to understand that the load could be moved at the needed acceleration rate with the given motor. What we were looking for, was to know the energy required in order to drive the motor.

Mathematics is a wonderful language, but math without practical application might not be able to fully describe all of science. What our mathematical models appear to indicate, when pushed to extremes, is not necessarily based in reality.

We have seen this blindness before in the past, and we will likely see it again in the future. And sadly, science is deeply immersed in it today.

Discrepancies are common when dealing with observations of scientific methods, and unless we can deal with these initial discrepancies, we will trap ourselves in our errors. Only a relatively short time ago, in the sixteenth through present centuries, many people in the Church were blinded with blind faith of religious assumptions. These assumptions ignored the discrepancies found in their own interpretations of the text of the Bible. This caused many Church leaders and people of the community to ignore the obvious scientific revelations that were presented to them by the likes of Galileo, and it is this blindness that needs to be avoided.

Galileo's and Other Great Thinker's Problem

Centuries ago, in the era when Galileo lived and the church reigned supreme, the people were sometimes pressured into compliance by political and church leaders. During the Renaissance Period some key families had great control over the church and politics. And while many great things happened through those people, the politics of the church were mostly run according to the wishes of those few families. At times, these families disregarded the best interest of the community and humanity overall. During that period there was much money being bestowed, by the church, for the purpose of community creativity. This was a good time for many of those who sought explorative, creative, or artistic types of employment. As culture shifted over the years, this new allowance for freedom of expression in artistic and scientific development opened the door for free-thinking and became a threat to the church's controlling parties.

When Galileo was told of a tool that would allow seeing things at a distance, as if those things were near, he decided to point the device toward the Heavens where he found a far more amazing and wonderful Universe than he had ever imagined. After he recorded his observations, he attempted to share those observations with the public and the church, but he met with resistance. He was asked by the church to recant his discoveries if he wanted to be able to continue his work—which was mostly funded by the church. Either his benefactors did not agree with his findings, or they were afraid and sought to stifle his message to the world.

Galileo's problem was nothing new, and it is also nothing old. The same issue that Galileo confronted so long ago is still confronting us today, and it will likely always confront us. It is this problem that we must avoid—it is a problem with humanity—it is ignorance and blindness!

Do You Suffer from
Scientific, Religious, and Evolutionary Blindness?

Our blindnesses and expectations differ somewhat between science, religion, and big bang evolution. Since science is an ongoing experiment, we should expect a certain amount of failure or let-down within that realm. However, with religion we fully trust, and then when our religion does not deliver as we believe was promised, we feel violated. We feel violated because of our incorrect understanding of religion and the accuracy of *who* we choose listen to.

It doesn't matter if our blindness is religious, scientific, or evolutionary; the blindnesses are all the same, because if we cannot see, then we cannot see. The likes of Copernicus, Galileo, and Newton are little different from the rest of humanity. While they each helped to advance our understanding of the heavens, or space, these great thinkers each had their own blind-spots— blind-spots that held them back from further understanding the specifics of what they each sought to know. They would likely have overcome them if they had lived long enough.

Most revolutionary advancements in our understanding of the Universe were fed by a deep desire to better understand the Creator and the Heavens. Even Charles Darwin was a religious man to an extent. He had both a scientific and a religious blind spot where he made the assumption that what he "knew" about religion was what *is*. His own personal religious blindness hindered him from more fully understanding his own intriguing observations—even unto his death. Darwin even made admissions that he wanted to be wrong in his observations of nature with regard to his understanding of his religion. Darwin had two blind-spots, and these blind-spots have become common among humanity, especially within modern-day science.

We believe that the Bible tells us that creation all happened in six short days, but our scientific calculations indicate that it all started billions of years ago. Religion seems to be increasingly rejected by people as each generation embraces the idea of evolution. Yet, even in the sciences there is a vast amount of disagreement, especially with regard to the "big bang" theory and the actual age of the Universe.

We are Being Betrayed

As humans, we have expectations of those who lead us—we expect them to be correct! Betrayal is a nasty seed that the Church has sometimes sown all too abundantly. If our religious leaders say that something is so, then we expect it to be so. When we find that the interpretations of the Church leaders are wrong, in error, or false, then we feel betrayed. When our leaders are not correct we tend to do one of three actions: We embrace their wrongness, or we reject everything that they say and we rebel, or withdraw altogether and choose to not care at all. However, the betrayal that we feel about the inaccurate information given to us, is our own fault for putting our trust in the wrong people. By wrong people I do not mean "religious people." I mean that we trust others to be perfect when they are not; this includes clergy, scientists, college professors, and anyone who we choose to make as our mentor. The expectations of anyone or anything that each one of us has, are critical to our ability to understand anything.

What we call "*science*" has brought human thinking to a new and better place than did some of the religion of our past. However, in the same way that we question and doubt some of the proposals expressed in religion, so too, when new scientific evidence arises, we should also be skeptical of it. If we immediately believed every new theory that came our way, then we would constantly be wavering and changing our conclusions, daily! This does not mean that we should refuse to review the new information in its proper time; but rather, we should approach it *without* an agenda to discredit it or without an agenda to blindly support it. In the end, the truth will always prevail when we are not deliberately choosing to be blind to it.

Just about every time that there is a new theory proposed in science, we reject it because we have a religious-like belief in what we have previously chosen to accept and believe as fact.

When enough evidence is finally presented to us, we come to an undeniable position that would make a fool of us if we did not promptly adjust our thinking.

In the world of scientific discovery this problem is very pronounced because scientists often pour their entire being into their own theories, and then they work to force what we believe we see around us into compliance with that theory. Each scientist's reputation hangs on his or her own theories; and if their theory is disproved, then that scientist's reputation typically goes down with the theory.

Consider Einstein's famous equation $e=mc^2$: this is a very assumptious statement, but it is a statement that appears to explain the relationship between energy, mass, and a constant. The "c" constant we use is humanity's scientific interpretation of the speed of light. Often, adjustments are used in this simple equation in order to compensate for its various discrepancies, but its principle essence appears to remain the same. As a scientific society, we are assuming that light is constant when there is no actual absolute proof that this is so. Einstein's simple little formula has rocked the world, and, because of it, he is regarded as the foremost genius of his century. Many of the things that Einstein theorized have been taken as fact; however, this does not mean that everything he said was correct or factual. Based upon his general thoughts and philosophies, it is likely that Einstein would not be fond of what we do today in science. This is with regard to our belief in the absoluteness of potentially-flawed current theories, and our frequent refusal to entertain solid new theories.

Do You Understand the Properties of Light?

The unique nature of what we call light, is a critical topic with regard to physics. Since light seems to behave more reliably and consistently than anything else that we are currently able to detect, it makes for a suitable index for measuring a multitude of aspects of our physical environment.

If light did not have its consistent property, then what we experience with regard to light could be drastically different. The instantaneous nature of light is a very good thing for us. Sound travels rapidly, but light is many powers faster. If light traveled at the same speed that sound travels at, then it would cause us a great deal of trouble. Our judgments about our immediate physical environment are based upon what we experience, and light has a great deal to do with our judgment on timing.

We generally do not realize it, but we expect sounds to come to us *after* they have originated. An enormous amount of how we react to sound is based upon its speed. Sound travels at different speeds with different air densities and different air temperatures, but, for all practical purposes, those variations do not cause us problems. If light, however, were to move at the same speed as sound does, then we would have difficulty accomplishing many common tasks. The problems would increase as the distance increased, similar to the delay we experience in the echo of sound.

Take, for instance, the delays experienced while controlling a Mars Rover. It's no small task to remotely operate a Rover that sits on Mars, here from Earth. With the delay of many minutes you cannot make rash moves when it takes minutes to respond. Imagine driving your car to the local market: What would happen if every time you made a move it took many minutes for your car to respond to you, and then a similar amount of time for you to know if the car did as expected? You would have to make precise calculations and then steer, brake, and accelerate in advance, all while not knowing for certain if your decisions were good until minutes *after* your car responded. Having to plan your trip to the market in this way would be a very difficult task. You would need to know the exact speed that you are traveling at, and the exact rate at which you accelerate and decelerate. Every decision that you make would have to be made with precision and executed minutes before needed.

If light and radio waves traveled at the same speed that sound does, we would have to send a command years in advance to Mars, and then wait years for the feedback to return to us from a Rover vehicle. The further away something is, then the greater the room for error. This means the greater the error— then the greater the error. What I mean by "the greater the error—then the greater the error" is that if we are wrong and we base our calculations upon our wrongness, then the error will often increase as the error increases in a somewhat exponential manner. Error causes more error, and can compound very quickly.

Light appears to have a consistent nature to it, a nature which has a stability that appears to us as unchanging. This apparent constant nature is what allows us to accurately control a machine on Mars millions of miles away. If light was very inconsistent in its speed it would cause problems in our calculations of when to make a Mars Rover deploy its landing preparations, and start, stop, or turn while traversing the distant planet.

In science we believe that light is unchanging in speed, and that from an observer's perspective it is always going to be moving at the rate of approximately 186,000 miles per second. In fact, we depend so heavily on this, that light is the standard for most scientific measurements, especially distance.

Light's speed is, without question, stable enough for us to use it effectively for our day-to-day lives and general scientific study. We have gone so far as to say that no matter what, light's speed stays the same for the observer; and then to compensate for the variations and anomalies that we think we see, we choose to mathematically change distance or compress "space-time." Our alteration of space-time works reasonably well, because it is a part of the mathematical calculations that are needed to calculate proper timing for landing a remote spacecraft.

Within its scope, when one number is changed, then the other numbers become larger or smaller as a percentage of the changed number, which makes the calculations *appear* to be *mathematically* correct. Any inaccurate data that is to be multiplied will have its errors reflected in the mathematical outcome at the multiplication factor used in the equation.

We misunderstand the potential variability of light because light is almost instantaneous in most day-to-day circumstances. Yet we know that it takes time for light to get from one point to the next. For instance, it takes about 8 minutes for light to get from the Sun to the Earth, and for light to go from the Earth to the Moon and back, it only takes about 2 ½ seconds.

The Earth and the Moon travel at basically the same speed around the Sun, which is a bit under 70,000 miles per hour. (Though, the Moon's speed relative to the Sun does change depending upon its position relative to the Earth while orbiting the Earth as they orbit the Sun together.) This means that if a light beam was shot from the Earth to the Moon and back, that in the 2 ½ seconds it takes the light to get back to the Earth from the Moon reflector, the Earth would be roughly 50 miles out of position when the reflected light finally returns. We are either compensating the 50 miles of shift, or the light is actually moving with the Earth relative to it and the Moon. It's difficult to detect this sort of thing because the slightest misalignment in trajectory will cause the reflected light to be off thousands of miles. And any light received back is very widely scattered and difficult to detect to begin with. When trying such distant experiments, any small inconsistency or variability will cause an enormous change in results.

Is the Universe Expanding?

Much of the "big bang" expansion theory is based upon our interpretation of light and its apparent speed, but there is a great deal of controversy in the scientific community with regard to an "expanding Universe."

Back in the Biblical Creation days, we believed the following controversial brief Biblical account:

"In the beginning God created Heaven and the Earth. And the Earth was without form, and void. Darkness was on the face of the deep. The Spirit of God moved on the face of the waters. God said, Let there be light. and there was light. God saw that the light was good. God divided the light from the darkness. God called the light Day, and the darkness he called Night. And the evening and the morning were the first day. And God said, Let there be a Firmament in the midst of the waters, and let it divide the waters from the waters. And God made the Firmament, and divided the waters which were under the Firmament from the waters which were above the Firmament, and it was so. And God called the Firmament Heaven. And the evening and the morning were the second day. And God said, let the waters under the Heaven be gathered together unto one place, and let the dry land appear, and it was so. God called the dry land Earth; and the gathering together of the waters God called Seas, and God saw that it was good. God said, Let the earth bring forth grass, the herb yielding seed, and the fruit tree yielding fruit after his kind, whose seed is in itself, upon the Earth, and it was so. The Earth brought forth grass, and herb yielding seed after his kind, and the tree yielding fruit, whose seed was in itself, after its kind. God saw that it was good. And the evening and the morning, the third day And God Said let there be lights made in the firmament of heaven, to divide the day and the night, and let them be for signs, and for seasons, and for days and years. To shine in the firmament of heaven, and to give light upon the earth. And it was so done. And God made the two great lights: a greater light to rule the day; and a lesser light to rule the night: and the stars. And he set them in the firmament of heaven to shine upon the earth. And to rule the day and the night, and to divide the light and the darkness. And God saw that it was good, and the evening and the morning were the fourth day."

Around the fifteenth and sixteenth centuries the perception of this account of creation began to take a scientific turn and started to be challenged. When open-minded observers began to notice that their observations did not appear to match with the interpretation that the Church had of the Biblical Creation text, they found themselves often rejected and threatened, and they were demanded to recant their theories. Eventually their observations won out because once their theories were released or leaked out publicly, then other people could test those theories and clearly see that they were more accurate than the Church's previous interpretation of the Biblical account of Creation.

When Darwin came along and postulated long-age natural selection, his arguments for it were believed to be undeniably compelling. However, he did not designate the specific amount of time it took for the evolution process to occur. It was simply speculated that there was a prolonged period of years that allowed for natural selection to evolve primates. Slowly, long-age theory began to be accepted and it seized the minds of scientists world round. This caused a relentless pursuit to redefine the age of the Universe.

Based upon the speed of light and the "red-shift" seen in the Doppler Effect, many scientists believe that we can accurately calculate the age, movement, and distance of celestial objects. Edwin Hubble's assessment of his own observations have led us to believe that the age of the Universe is about 13.7 billion Earth-years old, at last count. The 13.7 billion years old age is based upon an expanding Universe theory. The most distant galaxy observed is believed to be slightly less at 13.2 billion light-years away, and it is believed to have taken 500 to 700 million years for that galaxy to form. To many scientists, Hubble's "red-shift" observations are evidence enough to "prove" that the Universe is expanding and that it is 13.7 billion years old. However, there are other scientists that are not convinced of an expanding Universe, or of the stated 13.7 billion year age. These other scientists are abundant, and their theories range from long-age 13.7 billion years, all the way down to some creation scientists who hold to a 6,000 year history since the point of Creation. The conclusions that most scientists make, do harbor a bias based upon his or her own chosen belief system. Most proposed theories are credible and worthy of debate, though there are some that are utterly absurd. In general, most publicly stated theories are well thought-out and are very thought-provoking.

It is made quite evident in the textbooks for school children and college students, and in many articles written for scientific journals, that the consensus of public thought in the western world is that our Universe is expanding, and that it is 13.7 billion years old. Any such information that is placed into these text books and other sources of information will influence the young minds reading the information, causing religious- and scientific-style blindness of potentially incorrect assumptions.

Is Our Model of the Universe Accurate?

Late in the twentieth century, our view of the Universe was greatly altered by speculation of additional universes existing. Our view was also altered by advances in computing power and computer modeling that was being done with the plentiful newly available computer processing power of the period. Believing that there are additional universes enables us to look beyond our own Universe and potentially see things that exist beyond our most vivid imaginings. At that time the enhanced computing power allowed science to create speculative models of how we believe the forces of physics may have brought all of creation into being.

There's a long way to go with regard to full understanding of our Universe, and it is likely that we will never completely achieve what we are looking for, but I believe we should always keep trying. Only several hundred years back, many scientists and other people chose to believe that only some thousands of stars and a few planets were in the heavens, and that the Earth was stationary. Then as devices, such as the telescope, were created, the "heavens" became far more all encompassing to us than we ever imagined possible.

The continuing efforts to describe what scientists have been observing has brought about speculations of "Black Holes," a multiverse, dark matter, event horizons, and many more seemingly unexplainable conclusions. Most of these speculations are based upon the mathematics associated with physics and Einstein's famous equation. These calculations lead us to believe that certain phenomena will *always* occur. When we develop a hypothesis that we deem worthy based upon our calculations, then we attempt to prove our hypothesis through observation. This is an admirable methodology that can be used in order to further our understanding of our stellar environment.

Without the consistency of our mathematics, there is little that we would accomplish, because, typically, we humans struggle to realistically imagine beyond what we physically see as evident, and we also struggle a great deal to see beyond what we have been taught. Our *perceived* model of our Universe has been slowly ever-changing over recorded human history. The slow progression of knowledge is based upon humanity's slow progression of understanding, but, often, our "understanding" has been wrong and in dire need of correction.

When studying distant cosmology and physics, even our math takes us to something that we have difficulty comprehending. Debates about the origin of our Universe are nothing new. In truth, philosophers, mystics, and the general populous have been asking these questions throughout recorded history. And as far as we have specific accurate information, "recorded history" includes the whole of human history.

Creation based upon a God, versus creation based upon natural occurrence, is not a new debate; though in millennia past it may have been described differently. During the nineteenth, twentieth, and the early twenty-first centuries, many scientists believed that there is no god, and that all of creation originated from an infinitely small point, when suddenly and spontaneously, the infinitely small point exploded with no decisive interaction. Subsequently, stellar systems and galaxies developed as a result of the forces created during this theoretical "big bang", and then over billions of years, bacteria developed and morphed into all of the creatures that we know today, including humans.

Consider this thought as you read: We designed a computer program to make a model in order to see what we think the Galaxy and the Universe will do over time, and we model how we believe it formed.

Does God Matter?

In centuries past, many of those who sought out the deep questions of the Universe had profound belief in a higher power. For most of these scientific pioneers who observed the heavens, their belief in their God was the driving force behind their research and speculations. Without the undying dedication to their God from these great minds, it is possible that, today, we would remain in the dark with regard to much of what we feel we "know" about our beginnings.

While many people have concluded that either there is a God, or that there is not a god, in reality, the debate has only just begun. The debate will likely continue until we are all dead and cease to exist, or until **G/g**od shows us otherwise, whichever happens to be true. This is not something that *we* get to choose, because what is—*is*. What we believe, or do not believe, will not stop what *is*. It does not matter what *we* believe, because God or no god—whatever **is**—will at some point become evident when we keep looking from different perspectives. There is nothing that we can do to alter this simple truth.

In accepting the reality of—*what is, is what* **Is**—we can get philosophical, or we can delve into the mathematics of theoretical quantum physics and imagine that we can change the "what is" part of the equation. But this then causes us to have to question what is God, if there is a god. For our purposes of the discussion about origins of the cosmos, or the heavens, it will be more scientifically appropriate that we refer to a God as a "Creator."

This book is mostly about the scientific observations that humanity has made during recorded history. However, since much of the origins of the Universe are speculative, and often press towards the metaphysical, there are times where a **G/g**od will be mentioned throughout this book in reference to the Church leaders' past inaccurate conclusions, and in reference to the relatively newer scientific views.

Referring to God as the "Creator" dispels much of the petty "religious" debate and focuses more upon the actual topic of "Creation." Whether "Creation" was accomplished by a supreme Creator, or as random occurrence through the specifically mathematic consequences of nature, is not for this book to say; but we will be answering some very deep questions that may shed some light for many people who are interested in the subjects of g/God, Creation, evolution, and big bang.

My goal is not to dictate to others what they can or cannot believe. My goal is to open the hearts and minds of science-minded people around the globe, so that we can stop believing not-true things that blind us from being able to advance our understanding of our beginnings.

Much of our blindness is created from our own biases that are due to our learning and upbringing. In general, the biases of humanity have held us back from advancing as is illustrated throughout recorded human history, and such biases will likely continue to do so. Our own potentially erred interpretations of what we see around us, blind us from being truly open to receiving new true information.

For instance, it's true that we have found a multitude of fossils, and that those fossils have been found to have a diversity that was unanticipated until recent centuries. Yet, while we have found the fossils, the actual age of those fossils is not specifically known because the methods of proving those ages is somewhat speculative. The estimates that we have made in regard to the ages of these fossils are based upon our previous assumptions, theories, and findings, as well as being based upon our modern conclusions.

Specifying the age of fossils, the age of the Earth, and the age of the Universe has been, is now, and will likely continue to be a point of contention between the various scientific and creation belief sets for a long time to come. This is both a problem and a blessing, because, as history proves, humanity has a tendency to make assumptions, and then we often hold those assumptions as a core belief unto our deaths, only to be proved wrong not long after our deaths.

Our assumptions and beliefs are always based upon our interpretation, or someone else's interpretation, of speculations and observations made by past inquiring minds. We form these conclusions based upon our own understanding. If our understanding is incorrect, then our interpretation of our observation will also carry the properties from our incorrect understanding; and thus, our observations and conclusions will likely be incorrect as well.

If we insist that there is a God, then our science will reflect that. And conversely, if we insist that there is not a god, then our science will reflect that. We see this in the debate of evolution versus Creation with regard to the arrival of mankind on this planet. Evolutionists typically insist that there is no god and that everything took billions of years to occur, where it is said that many Creationists typically believe that God made everything in six twenty-four hour days, and then took a little me-time on the seventh day.

Does God matter? Our belief or disbelief in a G/god greatly alters our interpretation of our findings; and more importantly, it alters what we actually look for in our research. For instance, with evolution, there are many people who have an agenda to disprove god and the six-day-creation theory. People who are searching for the "missing links" desperately seek to find their missing link in order to prove their theories. Then, in response, and in order to achieve their own agenda, some of the people who believe in God will try to discredit the diversity of the fossils that are found.

It is because of our biases that these battles of science have been raging on in various forms for thousands of years. And because we blind ourselves, due to our own agendas, we often fail to find the truth. This causes us to have to try to force our theory on others until we are each proven wrong and the truth reveals itself. If we believe that we have the answers to the actual origin of the Universe, then we had better guess again, because it is when we think we *know* we have something completely figured out that we stop learning. This is because we stop trying, and when we stop trying it causes us to overlook what is truly going on.

The Importance of Definition and Consistency in Science

The various science maths are a universal language that all people can use to communicate due to the consistency of those maths. Throughout the entire globe, no matter what language uses the equation, two plus two *always* equals four. Our many spoken languages, on the other hand, are considerably more obscure and require far more thought in order for everyone to arrive at the same scientific conclusion. This is because the meanings of the words vary considerably within a language, and those meanings vary even more so when translating between languages.

We'll be delving into the Universe, physics, time travel, gravity, and our origins throughout this book. However, it's a waste of the reader's time to discuss any of that before discussing the importance of—agreeing on the values—assigned to the terms used to describe our sciences. So, let's consider the value of accuracy in the definitions of our basic scientific vocabulary.

Chapter 2

Bending the Ruler

There is a great deal of confusion in people as they absorb the information being spread by the dominant portion of the scientific community. By simply listening to what they say, surprisingly, the closer the person is to the scientific community, then the more confusion we often seen in the person. While there is a risk of offending a large number of science-minded people in making that statement, there are many very good and impartial scientists who care nothing of ego and only seek the truth; it is these scientists that have been making wonderful contributions to the world of science for centuries!

The word *science* means "*knowledge*", or to "*cut*" or "*split*," indicating investigation techniques that bring about *knowledge*. It is the job of science to split hairs and truly understand what causes it all to be. However, it's our *inaccurate* speculations and, more importantly, our inaccurate conclusions, that get us into trouble. This is true for science, and for life in general.

Regardless of what the specifics are, what we call "science" is a fascinating field of work to be involved in, and many wonderful developments that have been of great use to humanity have come from science. We should continue our quest to understand, but we should do it with open minds and open hearts.

It is when our ego has become our agenda that we are willing to bend the ruler of science. What do I mean by *Bending The Ruler?* Many scientists are willing to alter reality, in their minds and in the minds of others, in order to force their data to work to their own expectations—in essence, making two plus two equal five, whenever needed, in order to "prove" their theory. Mathematically this is difficult to do, so the complexities of the calculations are often increased by several orders of magnitude in effort to compensate for our human errors in thinking. It's not easy to make numbers do truly random actions.

Often, what we believe is random is more an issue of timing. Timing is a most critical component of science. If we don't time our observations correctly during an experiment, then we won't be observing anything because we will miss our observation opportunity. Additionally, if we incorrectly calculate the timing of things past, then we will not find what we are looking for in the immediate moment or in the future. Timing goes far beyond what we think of as "time", and it touches every other index that we use to navigate life and science.

Importance of Accuracy in Definitions

The title of this book, *Bending The Ruler*, is about science indexes and how we use them. Everything in life is about measurement. Without an index we could not navigate life, and science would entirely cease to exist!

In the last chapter we discussed controlling a Rover exploration vehicle on Mars from the command center here on Earth, and we discussed the precise timing required due to the delay caused by the distance that the communication signals must travel. We also briefly discussed our blind expectation of the delays that we experience with sound while interacting with our own physical environment. Each of these delays that we experience is an issue of timing. When the distance is small, then the margin for error is negligible and the timing issues are essentially irrelevant.

In a case such as an interplanetary Rover vehicle investigating Mars, the margin for error changes greatly with distance and becomes a critical factor, and then much more care is needed in order to accomplish a given task. This seemingly obvious observation is not as obvious as we might think when addressing other areas of science.

Timing is a measurement of sorts. Think of an American football team where the quarterback throws a thirty-five yard pass to a moving receiver. Both the receiver and the quarterback must make calculations in order for the exchange of the ball to be successfully completed. The quarterback must estimate where the receiver will be when the ball reaches the receiver; and the receiver must estimate what speed he must run in order to be under the ball when it hits his arms *after* he has seen the quarterback launch the ball. In the case of football, the receiver has a great deal of control because he can make the needed adjustments to his calculations at each moment in time. The quarterback, however, has zero control once the ball leaves his hand. If the quarterback fails in his estimations, then the receiver will not be able to accomplish his task. No matter how good the receiver's calculations are, the pass will not be completed because the ball is too far off the intended destination point-of-intersection, so he will not be able to get there in time to catch the ball.

Our everyday lives are governed by our individual ability to time events. If we do not apply the brake at the proper time while driving, then we will not be able to stop our vehicle at the appropriate spot, which will likely cause an accident because we will end up stopped somewhere in the intersection.

Timing is important, but timing is not so much "timing" as it is a definition. Timing... a definition, you ask? Yes, timing is a definition of how long it will take something to occur. The definition of "how long" is what allows us to function from day to day. Did you ever miss timing on something as simple as turning the door knob far enough in time to get through the door? When you approach a door, with seemingly little thought you reach out to grasp the door knob, you turn it, and then you push the door open. There is a great deal of critical timing involved in this simple function, and if your perception or *definition* of time changes, then you will be in for an abrupt surprise. If the door knob does not function as smoothly as expected and causes a slight timing glitch, then it will cause *you* to crash into the door.

Discussing definitions, especially for something as particular as science, can be a cumbersome task because with each new bit of understanding, we can try to break each subsequent explanation down even further.

But first, we have to define "*definition*" in order to be able to define anything else. All of our *definitions* are a means to communicate, and this includes our mathematic *definitions*. *Defining*—sets a boundary or limit on what we will allow something to mean as we share our own thoughts with each other and the world. Our definition for something is a *standard* that we agree upon, and we use these standards in order to effectively communicate. Whether it is words or numbers, if we do not have the same index of definition, then we will be unable to communicate effectively, as is evident in many personal, social, and business relationships.

Bending The Ruler is something that we all do every day without realizing it; and when we bend the ruler, we trap ourselves in inaccuracies and our own deliberate lies. Many of our words are either of ancient origins or are derived from ancient origins. The ancient source words that we now use were based upon the ancients' understanding in their contemporary times.

Our definition of how long something takes is a *definition*—
it is assigning a value to something. Understanding this is critical
in science; but, increasingly, it seems that many in the scientific
community have been overlooking this simple principle and task.

The word "definition" is typically associated with the
definitions of words in a dictionary, but, in reality, definitions are
far broader than we tend to give thought to. So, as mentioned, we
need to be able to define what the word "definition" means to
even begin the discussion.

The etymology of the word *define* is "*de*" plus "*fine*" or
"away from the boundary" according to Webster. This should
indicate, away from ambiguity and with specificity. This means
that we all know that a word such as "*no*" means *no*, and that
there is no gray area surrounding that. You can think of it in
terms of computer assembly language of I/O, *yes* or *no*, *on* or *off*.

Additionally, we cannot avoid psychology when discussing
science and the importance of accuracy. Human psychology is
incredibly important with regard to science and our findings. Our
human desire to be accepted by our peers is so powerful that we
are often willing to lie in order to be accepted or to have our
ideas be accepted. And often, even if we don't realize that we are
lying or wrong, we will still demand that we be accepted. An
example would be "It's the way I said it is; I'm right and you're
wrong because *I* said so!" And many times we even deceive
ourselves while doing so. This is why we are always *Bending The
Ruler*.

If we find that we're wrong, then others may reject us. No
human wants to be rejected. Humanity's primary function is the
desire to be accepted and to be loved. Nothing damages us more
than having some thing or idea that we held dear, be refused or
rejected by others, especially by our peers. Some people handle
this type of rejection better than others. It is in this that we find
an enormous unseen problem with regard to *Bending The Ruler*
of science. When someone is *correct* about something and they

attempt to share their findings with the world, but the world rejects them, then they often retract their statements due to their fear of continued rejection. Thus, their brilliant discovery goes unnoticed by the world while another more assertive person will *demand* that people accept the assertive person's inaccurate theories as true.

People often try to dumb things down with philosophical ambiguity. Such as the common attitude of "if a tree falls and no one is there to hear it, does it still make noise?" That sort of statement is little different than saying, "if you are cut and are bleeding and no one is there to see you bleed, then are you really bleeding?" Realize that we are all in the same world dealing with the same questions and problems. This book is for minds that seek the truth and can face up to the realities of our being *here now* to actually find answers to the questions that we all seek, and for minds that seek solutions to the problems that we all face. We are all in this together and we all share one reality.

Does Space Exist?

Here's a philosophical thought for you to consider: Does space exist? And from a religious perspective, we can go even deeper and ask: Does space exist without a Creator?

"*Space*" is perhaps the most interesting aspect to study within the scientific realm. In many ways, understanding *space* has been the ultimate quest for science. Is space empty, or does space actually contain something? If it does contain something, then what does it contain? If it contains something, is it actually still "*space*"?

Intuitively, most people realize that when we speak of a "space" we are referring to "*void*", though we often do not describe it in that manner. For instance the riddle, "What is in a hole?" the answer, "Nothing! If the hole has something in it, then it would not be a hole." This simplistic view of a hole is a whole lot deeper than one would imagine. For instance, a hole in the ground is void of dirt, and when children ask the riddle "What's in a hole?" they are referring to the fact that the *dirt* is void in that location, and therefore the hole exists—or does it?

When speaking in scientific terms, the "hole" just mentioned contains all sorts of stuff, so it's not really empty, and thus, it could be argued that it is not a hole at all. A hole dug in the ground of the earth will contain air and light and radio waves, etc. In many ways, space can be thought of in terms of a hole. The hole is a void, but void of what? It is here that, once again, the term *definition* or *define* comes into play. We cannot just go ahead and define what is or is not in the hole, rather, we must first define the word "*hole*." Etymologically the term *hole* is from "*hollow*" or "*helan*" meaning to "*conceal*" or "*hell*." Thus the comparison of "*hole*" to "*space*" is not adequate.

Now, the term "space" itself needs to be defined. The best that the dictionary can do is to give us the closest analogies of "*room*" and "*area*," both of which ultimately indicate open barren land of great expanse. Digging too deeply here will catch us in petty philosophical debate; therefore we will come to the understanding that space means void or absent of something.

Since the expanse of space is void of readily tangible substance, except for the areas filled by the celestial bodies, then what is "space", and does space contain anything? Being semi-scientifically specific, "space" is utterly filled with all sorts of radiation. To illustrate this, consider light from our star, the Sun; as far as we can tell, the light from the Sun can be seen from anywhere in space where it is not obscured by another celestial body. This means that from our lonely little star, that we affectionately call the Sun, all of space is filled with its spectrum of radiation emissions. Now, add to that the radiation from countless other celestial bodies, and space is anything but "space."

Is it possible that we cannot detect some aspects of space, or that we cannot detect what is in space? And if this is so, then do those aspects actually exist? If something is in a different realm that is unknown to us, then can we say that it actually exists? Is it possible for something such as matter or particles to not have ever existed? These thoughts are different than the,—if a tree falls and no one is there to hear it, does it still make noise?— thought, because that thought questions whether or not the sound exists when no one is there to witness it. Where, on the other hand, when something is undetectable, we cannot witness it even if we try because it is *undetectable.*

How Can We Measure Existence?

In our modern attempts to better understand, we have been perplexing about atomic structures, and we have delved into quantum physics by building ever larger and more powerful particle-accelerator colliders. This is done in an attempt to see what the various parts of atoms are constructed of. The deeper we peer into the existence of matter, then the more it appears as if it is all made from nothing. Is this possible? Could everything be made from nothing?

Now, once again, we are caught in definitions because we have three important words that we need to define: *Existence*, *Matter*, and *Nothing*. We must come to a certain consensus about some of these most basic terms if we are going to intelligibly discuss any topic that references them.

From a Creator perspective, the Creator always was, is now, and will always be. From a scientific perspective, we currently believe that all matter started from a single infinitely small point that had no dimensions, and then suddenly, from that single point, everything burst forth and the Heavens or cosmos were formed over billions of years.

Here again, we have to define the word *Heavens.* Just because the word "*Heavens*" was the chosen word used in a particular text does not mean that we should not use it in science. *Heaven* simply means to "*heave*" or to "*lift up.*" *Heaven* indicates "*up*"—from our perspective. As far as we can tell, space has no up or down. Since we cannot easily navigate space in order to find a basis for an *up* or *down*, we will have to settle for our Earthly perspective and accept that we look up into *space* or the *Heavens*; that is to say, the *ups* is where we see all of the stars.

The definitions of the fundamental words used when discussing our origins have caused us a great deal of blindness over the centuries, but we don't seem to realize it; though we all seem to be able to loosely convey the actual thought behind each of the pertinent words.

The words **Existence** and **Exist** are somewhat ambiguous. We get into a great amount of trouble in science with regard to understanding *existence*. There is a sort of hidden truth in the word *Exist*, and that truth is that we are trying to define it with science. If our deliberately assembled matter, which we call our scientific instruments, are incapable of detecting something, then is that "*something*" that we are incapable of detecting actually *nothing*? And if so, then does it *exist*?

"**Nothing**" is two words: **no** and **thing**. To keep this short and as non-philosophical as possible, we will accept our current understanding of **no** to be "*not*" or "*void of*," and our understanding of **thing** as a "*tangible*" (touchable) aspect of our existence, since we are speaking about our origins.

The last word mentioned that we must have consensus on is "*matter*." **Matter** indicates "*physical*," or "*substance*"; and **substance** means to "*sub*"-"*stand*" or stand below. It seems fair and honest to take the term **matter** to include the lower or smaller portions of the things that are made firm, such as particles and atoms.

Our definitions here seem to have gained us little because each new level of understanding appears to need additional explanation. However, reasonable minds should be able to agree that, at this base level, the meanings, as just stated, are sufficient. The terms "*existence*" and "*exist*" are the single most important words to understand when discussing physics, but there is an unavoidable philosophical, religious, and scientific debate with regard to the terms *"exist"* and *"existence."* This is where the ruler gets truly bent by each of the three disciplines: philosophical, religious, and scientific. Additionally, each discipline adds ambiguity to these definitions through internal debates between each of their own constituents.

Maybe we can approach *existence* from the following perspective: If you have a thought, then does that thought exist? I don't mean that what we imagine in our mind is suddenly manifested into a concrete tangible item; but rather, does a thought itself actually exist? If we cannot come to a consensus on this issue, then we will have a very difficult time dealing with things such as "dark matter," or a more antiquated term such as "æther."

While we can measure our brain activity, it seems that our *thoughts* themselves are intangible. Ponder this question as you read on: If something cannot be detected by our instruments, then does it *exist*?

New Size Determinations

We suffer from our inability to come to consensus on our discoveries, and we often go to great lengths in order to make our math match our wants or needs.

Sometimes, we humans are so eager to make our numbers work that we will bend the ruler for the not-so-critical matters in life such as new size determinations in our clothing. Some clothing manufacturers have adjusted their ruler in their arbitrary size determinations for clothing in an effort to entice us to buy *their* brand of clothing. We can imagine that their reasoning is that we feel better if we think we are wearing a smaller size than we are actually wearing. Size bending is done in order for us to justify the things that we don't want to hear or believe about ourselves.

Science is little different than the clothing manufacturers are with regard to size determinations. It's easy to just say, "Oh science, pfffft! They're always changing things to make their numbers work." This is true, but many of those numbers have given us tremendous understanding and have brought about many useful inventions. But still, *Bending The Ruler* is something that we must take diligent care to avoid doing or we will blind ourselves from the truth of the way things truly are.

It's very important to understand that we can successfully work with **in**accurate numbers and still robustly succeed at something. However, our level of success is greatly dependent upon our needed standard of accuracy for the given task. Take carpentry for instance, the accuracy needed in carpentry is very low in comparison with precision machining. In many cases, a structure framing-carpenter can be off more than an eighth of an inch and there will be no negative consequences; the craftsmanship will not suffer and we will be pleased with the high quality of the work. But, if your automobile manufacturer were to be off over an eighth of an inch in measurement, then it is likely that your vehicle would not run at all, and your wheels would fall off of your car. In fact, there are few if any screws or bolts on a vehicle that would work as expected with tolerance anywhere near an eighth of an inch. Most bolts would fall out of their place if they had that large of a mechanical tolerance.

Our all-inclusive personal and societal standards are very important to us in this world. Each individual standard is a language or means of communication in itself. The primary reason that science is able to bend the ruler in the way that it does is that for most of the population it simply doesn't matter. Who cares if the furthest galaxy is approximately 13.2 billion light-years away? After all, it has no immediate impact on the general populous, plus we can't even see those galaxies with the naked eye.

If science is wrong about the age of the Universe it is of little consequence to the workers that serve us our fast food at the drive-thru. Yet, annually we spend **billions** of *their* tax dollars in attempts to answer these types of scientific questions. Therefore, in conducting the research, scientists should be honest, accurate, and *true* when accepting taxpayer funds extracted from the people who service the community by serving fast foods or doing any other available service for their fellow man.

Time Does Not Exist!

We never quite got to the base of the definition of what the word "**exist**" means. It would be good to be able to break **exist** into two basic camps: the religious-philosophical camp, and the scientific camp, but it's not that simple. If we were able to get away with that, then we could simply say that our *thoughts* exist when religiously speaking, but scientifically speaking there is no such thing as a "*thought.*" Breaking the term **exist** into two camps would be a very convenient but it will still leave us with a dilemma regarding quantum physics.

When we get to the point of asking the question, "What are the subatomic particles made of?" then the term *theoretical* quantum physics is often invoked. The deeper we look into what *matter* is made from, then the more we find that it's made from *nothing*. If everything that we see and understand around us is made from *nothing*, then are our *thoughts* also from that same "*nothing*"? In other words, do thoughts exist?

From a scientific perspective, we tend to say or believe that something does not exist if it is *in*tangible, and then we typically close our eyes to that which we cannot explain. Where, on the other hand, religion jumps into the God-mode and says "God did it all," and you had better not ever try to understand it or else— BAM!

Neither of the two perspectives, or methods, are helpful to the general populous, though, the two perspectives are helpful to one another. This is because those who vehemently propose their incorrect position, on either side of the discussion, will eventually be proved wrong by the opposing side. This reciprocal annihilation of erred theory will continue until truth is actually revealed. I believe we can avoid many such nonsensical arguments when we open up our minds and diligently work to agree on our indexes.

Discussing indices or indexes brings us to the subject of *time*; and "*time*" is one index where we don't just *bend the ruler* but we tie it in knots! Does time exist? You be the judge...

Physics mathematicians, such as Einstein, have been thinking upon the *time* issue for a very long time. Though, maybe the length of time—that we have been working on *understanding time*—depends upon how fast we are traveling. See if you can catch the errors in our scientific understanding of "*time*" as you continue to read.

There are many of us who would like to be able to travel through time; either going back to change our present, or going forward to see what to do in order to change our future.

When toying with the idea of backward time travel, we end up running into some paradoxes. If you were to go back in time, and somehow affect your past, then you might suddenly not exist today; therefore you would not be able to go back in time because you never existed because of the change you made to your past. I, myself, enjoy playing with these paradoxical thoughts on occasion, and I also enjoy entertainment along these scientific lines, but our musing at such paradoxes will not make something possible. When we speak of whether time travel is possible or not, we typically look to the mathematics used in physics for our answers.

Einstein's famous moving train example seems to indicate that time travel is possible to a certain extent: For instance, if two people are tossing a ball back and forth down the length inside of a boxcar on a moving train, the ball will be traveling at different speeds relative to someone standing on the ground outside of the moving train. When the ball is tossed with the direction of the train's movement, then the ball's speed will increase and will be the train's speed *plus* the ball's speed, but when the ball is tossed against the train's direction, then the ball's speed will be the train's speed *minus* the ball's speed. This, of course, is relative to the outside observer's perspective. However, to the people inside the train tossing the ball, the ball will be traveling at equal speed in both directions relative to them and the ball moving with the train.

What we observe will be occurring at slower rate from the traveler's perspective as we travel faster. With this fundamental concept in mind, if we travel fast enough, then we should be able to stop time altogether from the traveler's own perspective. However, I have heard people describe this same train scenario, but instead of the moving observer seeing things slow down, things would actually speed up. The varying perspectives with regard to Einstein's train example seems to be dependent upon how a particular person understands, or misunderstands, Einstein's example, and is affected by their own vision of the example's position of coming towards them or moving away from them.

It seems that Einstein's conclusions are based upon what is commonly called the Doppler Effect. In the Doppler Effect, as waves come towards you, the waves are moving at a frequency rate that includes the speed of the source emitting the waves. This added speed makes each wave pass by you faster causing more waves to pass in a given amount of time, causing a higher frequency. As the source object passes you, the waves pass by you at a slower rate because the source speed is subtracted from the sound wave's speed, thus altering the frequency coming towards you as the source moves away from you. We see this effect when a vehicle passes by us at a high rate of speed. As the vehicle approaches, the sound pitch is higher, and then as the vehicle passes us, the pitch of the sound emitted from the vehicle lowers because of the lower frequency of waves impacting our ears.

Einstein made what seems to be an accurate assumption when he stated that light waves behave in a similar way as sound does when an emitting sound source passes by an observer. This "Doppler Shift" in light and radio waves is the primary means that we utilize in various radar systems to make our measurements of speed and distance. As the body emitting or reflecting the waves passes by, there is a perceived wave speed change to the stationary observer. Many scientists have extrapolated this and concluded that *time* changes with speed, or that *space* is either *contracted* or *stretched*. The views on this vary widely.

Ever since "e=mc^2" has been proposed, we have come to believe that there is indeed a correlation between energy and mass. This is believed to mean that the more energy that we use to increase our speed, then the more the mass of the moving object will increase. Additionally, the more mass we have, then the more energy it will take to accelerate our speed. This has an exponential nature about it, thus causing an improbable infinite increase in mass and energy as the speed is increased. So, as it is commonly explained, as we move faster, time will slow down for us, and our mass will be increased.

In the equation e=mc^2, the speed of light is the "c" part; it is the Constant and it is about 186,000 miles per second. As we approach that speed, it is believed that time will slow down more and more, and the nearer we come to the speed of light, then the greater our mass will be. The theory implies that it will take an infinite amount of energy for an object to reach 186,000 miles per second in speed.

This is the point where we start to get into the interesting parts of physics and science. What is *time*? Does *time exist*? And, can we *time* travel? It is very important to understand that if we fail to properly define what "*time*" and "*existence*" are, then it is unlikely that we will ever truly be able to understand our Universe.

Time is a peculiar topic because few people, including scientists, seem to be able to wrap their minds around the fundamental subject of time. Time is something that we seem unable to manipulate to any satisfactory amount.

We believe that we see how gravity and speed effect time on our clocks in satellites. On a regular basis the agency that is in charge of the GPS (Global Positioning System) makes minor time adjustments to synchronize satellite clocks; this is believed to be needed due to the effects of gravity and speed on the time-keeping function of the satellites' clocks. The time-keeping devices on satellites are very sophisticated devices. The fact that the satellite clocks need adjustments different from identical clocks here on earth, we believe, indicates that something is causing their time to change. This need for correction speaks for the importance of the correlation between light, mass, energy, and gravity. With regard to satellites, it is important to note that the time adjustments that are made are extremely small. However, it is critical that these adjustments be made if we want our GPS systems to be accurate.

Depending upon the accuracy that we require for our tasks, we design our clocks to divide our time into slices of a given size. The accuracy of counting time from city to city became very important when things like trains came into existence. Timing became critical because if two trains traveling towards each other used the same set of tracks and the timing was off, then the two trains would collide if they were unaware that the other train was still in route towards them on the same tracks. At this point it was no longer acceptable for the various train stations' time systems to be off a few minutes.

The more humanity becomes connected, and the faster we do things, then the more important it is for our standard of time to be the same and have it be more finely divided into higher resolutions. This is true for measuring time and for measuring words, or better stated, *defining* time and *defining* words. So let's revisit the question: Does *time* exist? To better understand, let's dig into the word *time*: In the dictionary *time* is associated with the word "*tide*" and possibly to the Greek word for "*divide*." But what is *time*?

Can we *time* travel? Can we alter this "*time*" thing, or are we stuck with what we have?

Science seems to have found many ways to theoretically bend time and alter our existence. However, this does not mean that we are certain that we *can* time travel.

There's a reason that we have not in the past been able to, and currently cannot, travel time; and it is because "*time*" does not exist—not the way we think of it anyway!

What Does Exist Mean?

While we touched on this twice before; the term *exist* is too important to the discussion to not dig a little bit deeper since we are discussing *how it all came to be*. As indicated in the dictionary, the word *exist* means "*ex-*" plus "*stand, stop,* or *stare*." Based upon Webster's Collegiate Dictionary the term *stare* indicates "*solid*" or "*stiffen*." And the *ex* part means "*out of*" or "*from*." So, it's safe to say, that to draw the conclusion for something to "*exist*" scientifically, it must be *stable* and *solid*, that is to say that it must be *tangible*. We can experience *tangible* things in many ways.

Some things are readily tangible and we can immediately touch and feel them, like our physical bodies. There are also other tangible things that seem somewhat less tangible, like wind and heat. However, just because things like wind and heat seem less tangible, does not mean that they are any less tangible in a physics sense. We have many ways of utilizing the nature of wind and air, such as for flying, and we utilize heat for heating our homes on cold days. From a physical standpoint we know that these sorts of intangible-tangibles exist because we can control them and we interact with them in an immediate and real sense.

Our scientific analysis tells us that anything that has energy or mass, has mass or energy, respectively; and if it has either of those, then it is tangible; and therefore, based upon Webster's definition it *"exists."*

From a scientific perspective, the question, "Does time exist?" must follow a specific definition, and for the purpose of this book we will use Webster's version of the idea of **exist** which indicates solid tangible matter, which includes the electromagnetic spectrum. However, the idea of what **time** is, still remains elusive.

We measure *time* in many ways, and one way is to use a mechanical clock. Another is to use "light-wave" speed and what we know about it in order to measure time. We also utilize atoms to measure time in atomic clocks.

Time measurement is done in the same way we would use various measuring devices to measure the size of a room that we are planning to carpet. We find a device or method that best suits our needs based upon the scale and units required, and then we arrive at a count of units of the chosen device or method scale. This unit count can then be transferred to the given task; in this case it is fitting a carpet to the room.

Time does not exist in a tangible manner, yet we somehow are able to measure it and speculate the time it takes for the light from a distant star to reach us. In the case of a distant star, we use the measurement of the distance that light travels in a year, which is called a *light-year*. However, this frequent error in thinking is not a measure of time, but rather a measure of distance. Time is still measured in increments of *years* in this case.

What is Existence?

Exist and *existence* obviously share the same root, but *existence* is more of an action or state, as indicated by the "*ence*" suffix. With *existence* we more specifically speak of an active participation in what we detect around us.

In a religious mindset we can ask, "Does God exist?" Since we cannot specifically detect God in a scientific manner with some sort of God-detector, then based upon Webster's definition, we can say that God does not exist. That statement is music to the ears of every ardent evolutionist and big-banger world-round. But, I feel quite certain that we could gather several billon people who would disagree with that and demand a redefinition of the term "*exist*" or its more active counterpart "*existence*." As for claims of intangible ideas, the same holds true for scientific ideas; just because we theorize something, does not mean that it exists.

Since God does not exist in a scientific sense, we must then ask, is there a Creator? This is not a question that is typically greeted with a smile in the religious community. Conversely, in many cases there is an outright hostility at mere mention of a "God" when in the presence of some science-minded people.

Religious-minded people also tend to get upset when discussing the findings that appear to support long-age evolution, or even when someone begins to question the existence of God. As you may recall, from earlier, we committed to use the term Creator in an effort to not get tangled in petty philosophical debate. Let's imagine for a moment that the "Creator" exists and that the Creator Created all of Creation; but what then is the *Creator*? Is the Creator light? Particles? Forces? Or an old gray-haired man on a throne somewhere in space? What exactly is it that consistently becomes a contentious topic of debate when discussing the origins of the Universe?

Does Space Exist Without (G)god?

This book is generally about science's *Bending The Ruler*, but it's not only science that bends the ruler. As we discussed earlier, in order to ease our own consciences, we even bend the ruler in our clothing size determinations. Religion also bends the ruler. It is something that most of us are guilty of in some part of our lives.

We have seen religious ruler-bending co-mingled with scientific ruler-bending in great minds such as Copernicus and Galileo. There is also evident religious ruler-bending within the various religions themselves. Take for instance, Unitarians versus Trinitarians, Christianity versus Judaism; and the differences between various Lutheran factions, whose ruler do we use to make our measurements and determination of accuracy? Is there one God or are there three in the godhead? If our World's *basic* math had such vast discrepancies, then little would get accomplished anywhere in the world.

Coming to a consensus on the Creator issue would be good if we could determine the truth, but it seems that a consensus on the Creator may be a ways off. This is most likely due to the ardent nature of the proponents on either side of the argument.

For the sake of discussion, and for now, let us assume that a Creator exists. With that said, does *space* exist without a Creator? Or for that matter, we can also wonder, did *space* exist before the big bang if there was not a creator?

Space

The concept of space can be a bit fickle to ponder. Is it "space" if it contains any discernable activity or matter? We can't consider space identical to a hole because, with a hole, the implication is that the void is formed as a result of the removal of a portion of the surface or body in which the hole resides. However, with space, we do not understand exactly what space is void of, yet we grasp the fundamental essence of the thought of empty space. *"Empty space"* is a redundant term, and it is used by many of us in describing an empty room for instance. Most people grasp the fundamental thought of a *space* being *void*. It's when we get into science that we quickly realize that *"space"* is anything but empty.

For the purposes of this discussion, when referring to *space*, we are specifically speaking of an all-encompassing location that happens to be full of many wondrous contents. Think of our discussion of *"space"* as a room decorated and filled with furniture and other personal belongings. The space in a room is a place, and it has location; where *"outer space"*, on the other hand, has location that happens to be everywhere. Similar to the furniture and lamps in the space of a room in our homes, *outer space* is partially filled with the celestial bodies and their apparent radiation. For the purpose of this book, space is a giant, infinite, endless room, and it has no bounds, no walls, no end; just infinite boundless room or space to put stuff. This brings us to the origins of *space*. Not what is *in* space, but rather, the actual void itself or not itself.

Did or Does Space Exist Before the Beginning?

The following question must be pondered: *Did or Does Space Exist Before the Beginning?* Meaning, did space exist before Creation or the big bang? Whether it be religious or scientific, the terminology being your own choice, but the fundamental principle of the question remains the same. *"Did or Does Space Exist Before the Beginning?"*, whatever that beginning is or was?

The implications of a void *existing* are vast and wide. From a Webster's definition perspective, *space*, as far as we can tell, does not exist because it is *nothing*. Was there a place, a vast location, before the big bang or Creation? If we cannot come to a consensus on these questions, then we are sunk both scientifically and religiously. If the void of space was created at the point of a big bang, then what was there before that? More space? If space does exist, then does it exist beyond the big bang's or Creation's extent? Is space a substance that formed at the moment of Creation? These questions are unavoidably philosophical.

People of religion, typically, but not always, indicate that it is wrong to question these topics of origin this deeply, and sometimes religious people refuse to use a ruler at all. Where science typically buries its head in the sand and ignores what it cannot explain and/or bends the ruler in attempts to explain its inaccuracies and inconsistencies. Ignoring a subject actually cuts the ruler short, thus the ruler is incapable of measuring the vastness of the subject.

The perceived paradoxes with regard to space having a beginning, or any sort of limit, confuse us because we have allowed ourselves to become bogged down in petty ruler-bending debates.

When we can't explain something, like our expanding waistlines, then we just change the numbers and pretend all is well instead of dealing with the actual problem. This sort of behavior serves us well provided that we all agree on the new standard, but it does nothing to improve our health. And, in the case of clothing size, all brands do not agree, which is something that women world round will attest to; size 4 is not the same with every brand. Thus, we are still left with some remaining unexplained troubles when we bend the ruler. Additionally, in the event that the clothing size "4" would actually all be identical with every brand of clothing, but the physical dimension are increased for all brands, then we are still left with the fact that our waistlines are larger than we would like them to be.

The discussion of whether space existed before everything was created is far from over. This debate becomes immediately philosophical in nature, and is additionally obscured by our bent definitions. When we bend the ruler for the definitions of our scientific terms, then at some point, we become trapped in what amounts to lies.

Can we overcome this? That will depend upon our ability to accept new and accurate ideas when they are proposed to us.

What is Space?

In order to be able to be accurate in our thoughts on science and religion, there are some aspects of **everything** and **nothing** that are often addressed, about which we must come to a consensus.

Our words are even more important than our math is; therefore, we should give more thought to our words than we do to our math. But we do not do this. Instead, we bicker between ourselves and insist that math will answer every "science" question that there is; but math likely cannot answer everything. Our current math is good, but it falls short of complete explanation, and it is with our own math that we are truly *Bending The Ruler*. While the math itself is not necessarily bent, our definitions of what we are trying to describe with our math is not only bent, but in some cases it is severely mangled.

"*Space*" is one area that remains controversial, and various belief systems disagree and insist that their way is *the* way. We need to have these disagreements because we have a tendency to trap ourselves in our own inaccurate information when we run unopposed. With regard to people bending the ruler, in science this is happening more and more with each passing day, especially when we date the age of things ranging from fossils to the Universe.

"*Space*" cannot exist, not in a Webster's etymology sense anyway, yet *space* is there. The idea of "*Space*" either needs to have a new special word assigned to it, or we need to get very specific about the term "*space*."

My understanding of the term *space* is now and always has been—the empty void portion of an infinite and all-encompassing location. The best descriptive word that we have for *space* is one commonly used in many translations of the Bible—"*Expanse.*"

We can shake off any Biblical connotations and adopt the word "*Expanse*" to mean: an infinite, all-encompassing, undetectable, empty void, that is the infinite location where **all** things reside. We should use *Expanse* scientifically with this common understood meaning.

It could be considered a narrow view that the *expanse* always existed, or rather, always didn't exist, but it seems to be a strong observation that "*things*" are somewhere. It is that "*somewhere*" that we call the *Expanse*. This concept of *empty space* or better stated the **Expanse**, is undeniably obvious, and in general, we can all grasp the thought of *Expanse*, though the *infinite* part can be a stretch for many minds.

The *expanse* has a paradoxical nature about it because its concept is along the lines of limitless-zero. If it is limitless, then how can it be zero? In other words how can space or expanse be anything if it is nothing? The *Expanse* is not scientifically detectable because there is nothing to detect. It can only be detected in our understanding of the concepts of void and infinite. I believe that the empty *Expanse* always has been and always will be. It was not created and cannot be destroyed. It, when alone, is void of matter and has no end. We speak similarly about energy, but the energy resides within the expanse. Our belief that energy cannot be created or destroyed is only in our imagination.

A Foundation

Understanding the expanse is crucial; however, when discussing space there is an additional aspect that must be addressed, and that aspect is the *foundation* of things. Yet, there is another word that seems more appropriate than *foundation*. In most of our minds, the idea of a "**foundation**" indicates the bottom, and the *expanse* contains the *foundation* for everything. Because *foundation* implies a bottom, a more neutral word would be more appropriate; a word used in some translations of the Bible is "**Firmament**." If we err in understanding the *expanse*, then everything else that we think we understand will reflect our errors in this most base of topics. Again, I would like to toss out the religious implications associated with the word **Firmament** that are due to its use in various translations of the Bible, and drop any biases, and then utilize the word for its unique nature.

We can argue that the celestial bodies do not sit upon anything firm, but even discussing that view demonstrates a very narrow view. For instance, take water· It's a joy to navigate in; we swim, we jump in, we splash around, and we can immerse ourselves in it and interact with it within its space. We can even walk on water when we move fast enough, as is seen in bare footed water skiing, and if we hit the water too hard, then we find that it is very firm indeed!

The *firmament* of the expanse may have a similar function as water in that respect. All of the expanse may be filled with the *firmament*, but yet the *firmament* may not be detectable to us; and therefore, by definition, the *firmament* does not exist. Our atmosphere may be a better analogy for the *firmament*. We navigate through it, but in general we do not feel it, notice it, or pay much attention to it unless it is a part of our work.

What I am calling the *firmament* has been touched on in past times; and because we cannot detect it, we are unsure of what we are discussing. In the past it may have been referred to as "*æther*," or in the late twentieth century the term "*dark matter*" was used depending upon your view of those concepts. I personally have not spoken to those who initially declared those terms, therefore, I cannot be certain that their underlying thought was the same as my own thought with regard to me calling it the **firmament**. Since it appears undetectable, I prefer the term firmament because it allows for a broader view of what to look for when trying to detect the *firmament*.

This is not implying that other terms are all wrong; but it would be good for science and/or religion to be capable of having an intelligible discussion where everyone is grasping the basics of what is being discussed in this book.

If we all fail to understand something in a similar manner, then we will fail to be able to communicate successfully with each other. The Church has destroyed itself in this regard, and now it appears that science is doing the same. It may be just as well that this happens with science, because the scientific ruler is being bent so far out of shape that a good scientific shake-up is desperately needed.

Not defining words properly is far more dangerous than thinking that two plus two equals five. If someone were to come along and make a bold statement that two plus two is five, then they will be quickly caught in their error. They would be descended upon by every school child that can count cookies; and most likely, the perpetrator of the error would be ignored or mocked, for their ignorant statement, by the world and especially by first and second graders. On the other hand, words are considerably more abstract. Without clear and distinct understanding of the basest words used in science, ridiculously speculative statements will continue to be made, and will increase in absurdity as far as we allow.

We see outrageous claims being made on a regular basis that are similar to the flat Earth theory, but these often lack any logical credibility and are eventually disregarded when the error is quickly made evident. My concern is that these types of outrageous claims, hypotheses, and theories are becoming the norm, and that finding the truth will become irrelevant to people.

Truth is another word that needs to be scrutinized. *Truth* must be defined as what *is*. Here again, we can get philosophical and start discussing "your truth" or "my truth" and waste vast amounts of our time and energy. What *is*—is what *is*—and nothing that we do or observe will ever change that simple truth. It should always be the quest of science to find truth—without being scientifically or religiously biased.

The idea of "*truth*" is often associated with religion or Biblical discussion, but *truth* is universal. And while we can change things, we cannot change the truth about things. We cannot own truth, because truth owns us. Those who do not submit to the truth will ultimately be crushed by the undeniable power of truth, and then their own stubborn ways will become their destruction. They will be proven wrong at some point, and, in that day, their stubbornness will be made plain for all the world to see. This applies to all aspects of life—religion and science included.

Definitions Recap

Discussing definitions, especially for something as specific as science, can be a cumbersome task. With each new bit of understanding, we can attempt to break down each explanation even further.

We have to agree on the term "*definition*" spoken of earlier, in order to be able to *define* anything else. All of our definitions are our means to communicate, and this includes our mathematic definitions. *Defining*—sets a boundary or limit on what we will allow something to mean as we share our thought. Our *definition* for something is a standard that we all agree upon. We use these standards in order to effectively communicate our thoughts and ideas to one another. Whether it is words or numbers, if we do not have the same index of definition, then we will be unable to communicate effectively. And, as mentioned earlier, lack of index is evident in many human relationships—and it's worth pointing out again to emphasize the importance of understanding the *importance of index*.

The weights and measures agencies for the various governments are very particular about the definition of their weights and measures. This is so important that they have index weights that are kept in a protected environment in order to assure a reliable and constant index. If we would take such care with our words in regard to science, then we could more readily learn *more*, and we could do it with greater accuracy.

Chapter 3

High Expectations

When discussing the deeper aspects of *discussing space*, such as "What is space?", our own personal perspectives are incredibly important to account for; it seems that few people can successfully pull off an unbiased analysis of the data.

For instance, let's take the Earth-centric view of the Heavens: We believe we have good evidence that the Earth is not the center of the solar system, and that the Earth orbits our Sun; but let's not get arrogant thinking that we know it all; because, prior to Copernicus we had an Earth-centric view because of *our* perspective on the word *up*. And until Galileo introduced the idea of using the telescope for observing the celestial bodies, humanity only had the naked eye with which to observe the heavens. When we look *up*, "*up*" is a reference between our feet upon the ground, and our head in opposition to our feet upon the Earth. Right now, at the very moment that you are reading this, someone on the opposite side of the globe is looking *down* towards your *up*. While humanity has a good grasp that the Earth orbits the Sun, we are still susceptible to our own perspectives and biases. Ancient researchers (scientists) did a pretty stellar job at mapping out our solar system when we consider the resources they had available to use at that time.

If we truly desire to find the truth about our celestial environment, then it is important to firmly grasp *us* relative to the *environment*, rather than the *environment* relative to *us*. Many scientists get stuck on their own reference-views during discussion; and then not only do their views include our physical and directional references, but more critically, the views include, and also overlook, their own mental and *thought* reference biases.

I feel safe in saying, that in a few hundred years, people will look back at many of our current scientific assessments and wonder how we could have seen things so incorrectly. When this occurs, it will be similar to the way that we now feel about the Ptolemaic Earth-centric view of the heavens.

The Ancients

To demonstrate how basing our perspective on ourselves, rather than on our surroundings, becomes problematic, consider that it is believed that some ancients may have thought that they had to feed the Sun human blood-sacrifices to keep the sun alive. Imagine in our modern world thinking that we need to kill a human and offer that life to the Sun-god in order to keep the Sun lit—Woe to everyone on a cloudy day! We like to imagine that we have "evolved" so far that this Sun-sacrifice could never happen in our world, or at least in western society, but I would not be so sure of that. "Civilized" societies have come and gone many times over the thousands of years of recorded human history. When a society is past and gone, then the remnant persons of that society often adopt peculiar new behaviors as the new society progresses, and some of these new behaviors evolve into some very unbecoming human attributes as was repeatedly demonstrated throughout history. We humans are not safe from our own folly—ever! We must be ever vigilant in order to **not** become arrogant and imagine that we will not repeat the errors of our ancient ancestors. In the end, it was always the arrogance of society that crushed each society; and in its current state, this is where science is headed. There are too many people who have invested their entire identity in wrong hypotheses and now they cannot escape without being destroyed and stripped of their status. The science world seems little concerned of crushing and destroying *new* thought-provoking theories with what is believed to be the solid "proof" of the *current* theories. The science world is also notorious for attempting to wrongly crush true, proper, and articulate ideas that don't fit with the current contemporary scientific consensus. If you doubt this, then try to inject a new and *different* hypothesis into the scientific culture, and then feel the wrath of arrogance that so many before you have felt.

At any contemporary time, we are all just as vulnerable to error as the ancients were; whether those ancients were twenty-five hundred years ago, two hundred and fifty years ago, twenty-five years ago, or two a half years ago, it is our ignorance and our arrogance that traps and blinds all of us from seeing the truth.

Incorrectly Teaching New Minds

Often, with regard to science, the people who teach it to our children simply do not understand it well enough to accurately convey the intention of the original discoverer's explanation. For instance, in the double-slit light experiment a light is directed at a panel with two narrow side by side slits cut out of a panel. The light that goes through the slits spreads, from where it passes through, to when it hits the wall on the other side. As the light shines through the slits, the beam that shines through each slit overlaps each other and is seen on the wall. When it overlaps, it has a pattern of lighter and darker bars (Figure 1 Double-Slit Experiment, Page 50.) as it is often depicted. The fact that these lighter and darker regions occur is not in question. What is in question is why or how it occurs, *and* what it means to us. It is believed that *waves* cause this phenomenon, which is where the waves idea is supposedly "proven."

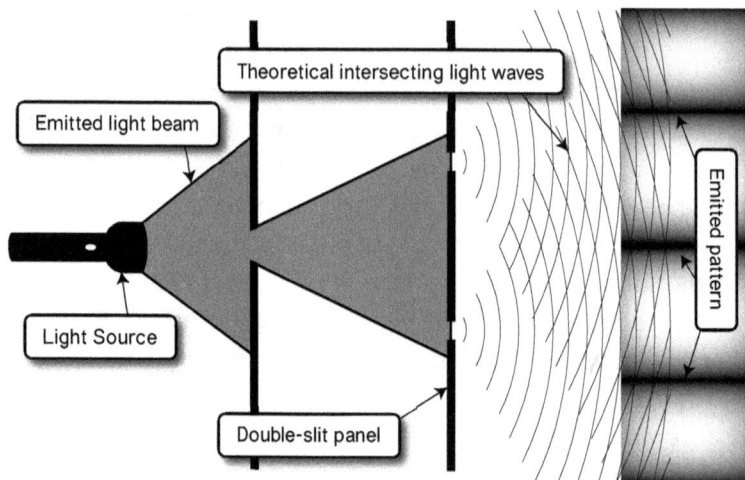

Figure 1 Double-Slit Experiment

The way that the wave/particle interaction of light is taught to children is perplexing. It is taught is as if it is *absolute* and that there is no other possible explanation. Yet, we have never actually seen the waves from light. Though, we do have some seemingly compelling evidence that light travels in waves and that light can somehow also be a particle. Additionally, the theoretical state change from wave to particle that is thought to be witnessed when doing the famous double-slit experiment, is being taught that this "proves" that a particle can be in two places at one time, *and* at neither place, *and* at either place.

We should have no problem with scientists making such speculations of a single particle allegedly being in two places at the same time, but we should take issue with the absoluteness with which those, yet to be proven, speculations are delivered to our youth.

Quantum physics is thought to be tricky and somewhat unexplainable, but this sort of thinking with regard to light particles, light waves, and quantum physics is no different than flat-Earth and Earth-centric viewpoints. Asking our children to believe such nonsense is much like saying, you **must** believe that God exists, or that long-age evolution is absolute, or that Santa Claus is alive and well today. In the end, most of what we believe about astro- and quantum physics is nothing more than fantasy-based speculation. Even the God of the Bible didn't want people to blindly follow. In the Biblical accounts, the Creator produced vast evidence of Creative power and used that as a reason to believe.

Eventually, we will find out why it looks as if a photon can be in two places at one time, but it won't happen anytime soon with a *belief* that it is absolutely in two places at one time, in order to stay objective about these types of questions we must stay clear of abstract thinking and seek to find out *why* the double-slit test appears to put photons in two places at the same time. We must consistently realize that just because an experiment *appears* to show a particular result, does not mean the apparent result is exactly as we perceive it to be. We must be able to *prove how* something is so; but, even when we can offer proof, it still comes down to what we accept as "*proof*" in each our own minds. All too often, our desire to prove our own theory gets in the way of reality.

Duality of Light

It is taught to students that light is both a wave and a particle, but there is an amount of uncertainty in this theory. This uncertainty is shown in experiments, with the double-slit experiment being the most prominent. In reading the numerous descriptions of this apparent duality, we can see that the interpretation does not necessarily agree from description to description. Some accounts indicate that light is both a wave and a particle at the same time, and others indicate a possible state change back and forth as required by the light. Logically, light appears to travel as a wave but transforms into particles upon impact/contact. As an analogy this can be thought of as similar to the state change of matter as is seen in solid, liquid, gas, and plasma.

Light does seem to have both a wave and a particle nature about it, but what if the true explanation is more beautiful and simple and more comprehensive than that? What if the particle observance anomaly seen in the double-slit experiment can be explained by simple, raw, hard, observable, and repeatable experimentation? Will we ever find the truth if we actually believe and teach the next generation with absolute conviction that the same particle can be in two places at one time? Not likely.

Patterns can be Explained

The "interference" pattern seen in the double-slit experiment is *assumed* to be *wave* interference. It is possible that this is the actual answer, but it also may not be. This radical statement might not be popular in the scientific community, but it is true that the conventional thinking on this may very well be in error. However, just because an *assessment* of an experiment's result is incorrect, does not mean that we cannot use the incorrect information to further future experiments or to improve all of our lives. You can think of this in terms of getting change back when you purchase something: If you were shortchanged by a few cents, the change that you did get back is still useful. Or we can think of the wrong assessment of an experiment as using the erosion from the water from a fire hose to dig a hole for a foundation for a new home. The fire hose will do the job but it can be done more efficiently with a backhoe or an excavator. If we calculate everything based upon an assumption that the fire hose is proper, then we can calculate how long it will take to erode the hole. We can calculate how many gallons of water are needed, and how much dirt will be removed. But, even though the calculations may be accurate, they still do not make a fire hose the right tool for the job. The fire hose may eventually get the job done, but it will make the job messy and create a great deal of extra and *unnecessary work*, and it will likely cause confusion as to why we cannot efficiently get the water out of the hole fast enough, if at all.

In the double-slit experiment we assume that light changes state because we have made assumptions that are akin to believing a fire hose is the best tool with which to dig a hole for a foundation of a house. When the light passes through the slits and creates the alternating pattern on the wall, we attribute it to wave interference, but it may not actually be waves causing the phenomenon that we witness. There may be another reason, but we will be hard pressed to find the true reason because most of the science world has not only accepted the effect seen in the double-slit experiment as wave interference, but has also celebrated it.

Any pattern can likely be explained by some process or mathematical equation. If a pattern is observed, then there will be characteristics of the pattern-producing substance that allows that pattern to occur. This is true in physics, quantum physics, and in biology as well. There will be characteristics in the cells, structures, organisms, molecules, atoms, etc... that cause the patterns. The characteristics of the substance are a *part* of the reason that the specific patterns occur.

In other words, if I want to make a pattern for something and I create the pattern, then the pattern will adopt characteristics from my manipulation *and* from the material that I created the pattern with. Then a subsequent observer can go forth and analyze the pattern I created to try to understand how it occurred, and they can explain why that pattern exists through mathematical reasoning, and through the observations of the complex portions of the pattern. However, they cannot mathematically explain *my* interaction in the process that led to those patterns.

Patterns are common and predictable, but we do not always recognize patterns; and sometimes when we recognize patterns we do not assess them correctly. For instance, if someone were to make fake, but very real-looking, footprints in the sand, we would likely assume that someone has walked there with their feet and made those prints. But, we would be wrong because the prints, in this example, were not made by feet, but rather only *resembled* markings from feet. In this example, the pattern of footprints has two components: First, there was a foot-shaped pattern made of wood, and then there was the pattern of impressions created by the interaction of a human hand repeatedly impressing the foot pattern into the sand.

Making incorrect assessments of patterns that we see is a danger, but until we truly have it all figured out, we have no choice but to make such assessments based on our observations and experiments. Yet, we forget that patterns can be caused by multiple factors that sometimes include outside interaction. Additionally, patterns can produce post-patterns, as can be observed in a moire effect when taking two window screens and angling the grid of one in relation to the grid of the other.

Any form can be added together and it will create a post form. These post forms can be added together and create a new level of post form in the same way that the building blocks of subatomic and atomic particles create all that we see.

Anytime we have made an incorrect assessment about the patterns that we see, then we have bent the ruler; this is true even if the incorrect assessment is done in innocence. For this reason we must remain vigilant in delivering the results of experiments in an open-ended manner by presenting our evidences, as *observed*, and not as *definite*. For example, I don't want the <u>speculation</u> of wave/particle state transformation; I want clear and concise observable evidence, because if we settle for anything less, then we have cheated ourselves and our children. Sadly, too many scientists have embraced what we currently *believe* that we "know" about wave/particle physics and ignore the fact that we simply do not "know." This is akin to using a match as a beacon on a foggy night for a ship that is searching for the harbor. The match is of little use and illustrates that the search to understand and see the light about *light* has only just begun.

Just because we *believe* we have found an explanation for a pattern does not mean that our explanation of that pattern is accurate or correct in any way. For instance, with light shining through two slits, we *assume* that the light travels in waves and produces wave interference resulting in a striped pattern; and it appears that there is good evidence to prove that. Water wave interference is often used as a suitable analogy, but the wavelength of water is far different than the supposed wavelength of light. This difference brings into question what we see occur in the double-slit experiment.

There is a type of person, whether it is in politics, religion, or science, who allows his or her own fears to turn to arrogance, and, in the realm of science, they subsequently desire to dictate to others what we can or cannot believe with regard to science. This attitude must be defeated or science will continue, even more so than it is now, to become the newly imposed religion that it has been becoming for centuries. This will be similar to what happened with the *forced* teaching of the human evolution principles to undiscerning school children. And this is absolutely no different than forcing religious teachings in schools. I'm not opposed to either being taught. The issues with these sorts of problems are not the actual teaching; but rather, the problem is the *forced* teaching. Using unwilling people's tax dollars for this sort of education is wrong because it promotes an agenda that has not been fully proven, as if it is factual, when it is not factual. The forced reading of it also rejects any opposing thoughts and opinions and refuses to allow *open* discussion, and when that is the case then "science" has fled.

The celestial bodies do not obey *our* commands, we only do our best, via mathematics, to describe their unexplainable and sometimes seemingly illogical movements. The only reason that we are able to mathematically describe the movements of the celestial bodies is because of the patterns that they display in their movements. Our interpretation of those movements often has been in gross error, but our flawed mathematic formulas allow us to be able to predict with a reasonable amount of accuracy where a body will be and when it will be there. However, this is more a result of repetitive observation than it is a result of accurate mathematics. Early astronomers could very accurately predict location by little more than repetitive observation that was eventually explained mathematically due to the consistent repeatability of celestial movement and accuracy of those early predictions.

It's interesting how we marvel at the accuracy of the astronomy of the ancients, because, as far as we can tell through archeology, they did not have as articulate of math that we have or the precision instruments that we have in our more contemporary times. Yet, our calculations are not much different than theirs. This makes it evident that, with good observation, it's far less important as to what tools we have, and it is more about *how* we use our minds.

Our math is a very useful tool, but it does not dictate the movements of the heavens. Rather, the movements of the heavens dictate our mathematical formulas. This seems obvious when we take a moment to reflect on this point, but in science we certainly do not behave as if we understand this simple truth.

Observed Patterns Become Laws

When we hear some of the talking heads of science, their words are often delivered in such a way that *we* (humanity) make the rules and set the laws. We believe that we have figured it out and state our case as a "law", and then we impress our laws onto the rest of the people.

We believe that *we* can control anything—we are self-centered. I'm uncertain who first laid down the term "*law*" when it comes to physics, but especially with regard to physics, the term "*law*" is terribly misguided; or, at least our current science world is misguided in its understanding of those "*laws.*"

When Newton laid down his "Laws of Physics", he was observing what he had seen, and he made some accurate observations based upon his interaction with his own environment. Newton went further and extrapolated from his observations that his laws also applied to interplanetary and interstellar actions.

From our contemporary observations, it appears that the celestial bodies *obey* Newton's Laws of Physics. Based upon these "laws," we have safely been able to land cargo on the Moon, Mars, and other celestial bodies, so Newton's laws have obviously proven to be invaluable to us. Any assault on those laws is met with a great deal of hostility from the scientific community; yet, I will say that his laws are not laws at all. In our human arrogance, *we* make the laws, and then we pretend that everything obeys these man-made scientific laws. If we cannot get people to obey our civil laws, then what makes us believe that the planets will obey *our* scientific laws—what arrogance! Personally, I do not believe that Newton's statements of laws were meant as many of us take the term "laws" to be used today. When Newton used the term "laws", I believe he meant the way things actually work, and it seems that most people grasp this. However, when we hear the term laws, we respond to it as if *we* somehow control the terms of the *Laws* of Physics. We do not control physics—we *describe* physics. And when we finally get close to accurate, then we establish our observations and call them "laws." It is our arrogance that somehow tells us that *we* make the rules of physics that all things follow, when in reality, we don't. We only make up "laws" based on our interpretation of what we believe we have observed, and then we demand our way by forcing or fooling everyone into believing that what we *think* we have discovered is true and that everything obeys our laws.

The public face of science, which is made up of the talking heads of the science community, is especially bad in this regard, and they often crush anyone who dares to challenge their beloved "laws." This is no different than the laws that were imposed onto the people by the leaders of the Church in centuries past, or modern governmental laws imposed on the citizens of the land at that time. Prominent Church and government leaders proceeded to crush any opposition.

Yes indeed, the science world has become our new church! If anyone contests this, then consider studying case law, for example, regarding teaching "Intelligent Design" in the public school system in the United States, and then try having a discussion about it with certain prominent scientists. I am not saying that so-called "Intelligent Design" should or should not be taught, but rather, I am simply pointing out the religious nature with which the scientific community (which includes government support of science) approaches the debate by imposing arbitrary laws onto the people, both in science and culture in general.

To recap this very important point, our "Laws of Physics" are nothing more than our feeble attempt at trying to describe the infinite nature of the Expanse. The sooner we understand this, then the sooner we get to move on to a new level of understanding of the cosmos by using our sciences. Our "laws" only describe a few very finite aspects of the heavens. "Laws" is a horrible word, with horrible implications, to use with regard to science. "Laws" are rule-making, which is something that humanity has been all too fond of doing; and these imposed laws have been a large part of the downfall of most past societies.

Galileo, Newton, and Einstein have only created the laws in the mind of man. In reality, it is *we* who obey the laws of the heavens, and nothing we do can defy that fundamental truth. It is time for new bright minds of science to rebel against the science "laws" and find the truth of *why*, so that we can straighten the bent ruler that was mangled by a few prominent scientists. Can a person who demands that their *flawed* theory be accepted and taught in schools, truly be considered to be a "scientist"?

Does Light Have Mass?

Throughout recorded history, "science" has been *Bending The Ruler*. We have little choice but to do so. This is because we do not truly understand what we see, and we are only doing our best to understand it all. I have no problem with science temporarily bending the ruler, but let us make it our mission to keep the ruler straight in the long run, and as straight as is possible in the meantime. We should not try to stop anyone from making speculations, because that would be like stopping Galileo from making his observations and conclusions. Yet, we must be cautious that we *do not* allow others to stop *us* from making new observations and new conclusions ourselves because of *their* pre-existing beliefs, which is where the problem arises. The modern scientific community is doing *exactly* what the Church did centuries ago to Galileo.

The views arising from particle physics and quantum physics of the late twentieth century, literally have the ruler of science twisted, contorted, and bent into knots. The equation $e=mc^2$ has brought about a great deal of speculation about time-travel, multiple dimensions, and warped space. The imaginings of early twenty-first century scientists and theorists have become all too fantasy-based, often referencing entertainment and movies as their frame of scientific reference. Some of them reference this entertainment as if these plays of entertainment fiction are authoritative sources of scientific information.

It has been believed by some people that light has no mass and that light is unaffected by gravity. Yet, during a solar eclipse as light passes by the moon we speculate that it is being bent. Then on the other hand, we say that light's speed does not change but rather space bends. The only "Uncertainty Principle" in science, is our uncertainty of our own conclusions of what we *imagine* that we are observing.

To defy the laws of science we could say that if light can be affected by gravity, then it must have mass; and if light has mass, then it cannot be constant. But we do not truly understand what "mass" is, or rather, what causes it. Therefore, saying that light has no mass is an ignorant statement.

The twentieth century has ushered in "black hole theory," where the gravitational pull of a celestial body is so strong that not even light can escape its pull. This seemingly fantastical view is only a stretch of our imagination if we believe light has god-like qualities. It's time to accept, in a concise and straightforward manner, that light is a *thing* and that it can be manipulated.

The sensationalized name "black hole" masks a simple truth about the underlying star and the light that may be unable to escape its "gravitational force." If we are at all accurate in our observations of what we believe to be black holes, then light *can* be affected by gravity, and light's speed can also be affected.

With regard to light and black holes, a dull black spot on a wall is a black hole because light does not escape it; so it should not be a stretch of our imagination to accept the concept of a black hole existing. Observations made in the early twenty-first century appear to indicate that "black holes" are common in the heavens.

The ramifications to scientific measurement are huge if light can be affected by the gravitational pull of a star. If we want to find the truth, then we must break through the error of our laws, and realize that our laws are only a finite description of what we witness. When we abide solely by the "Laws" that *we* invent, then we cannot see the other truths awaiting our discovery.

The problem of our potentially erred view of the properties of light cannot be completely laid on Einstein, and I do not want to affect his credibility. However, our assessment of his statements may be grossly flawed. Einstein bent time and space and left light constant. Since the light's speed is so great, our ability to measure that speed is left wanting, and it allows a tremendous margin for error that we can choose to utilize in order to make our wildly preposterous hypotheses. It is far less likely that the empty void of the expanse bends, shrinks, or changes in any way, than it is that light's speed changes.

Exactly how Far is the Farthest Known Galaxy?

We have no true means of judging distance of how far away the stars are from us, let alone the galaxies. We may be able to use triangulation, which is a logical and reasonably fair method for finding the distance of nearby stars. However, to imagine that we can come anywhere near understanding the age and distance of the farthest known galaxy with our current methodologies, is pushing the limits of reasonable thought.

There are a multitude of anomalies in our bases for gauging the speeds and distances of other galaxies. Since all of our assessments are based upon our belief that the speed of light is unchanging, any error in our assessment of the light will be transferred to our speculations of what we measure with that standard.

Given our contemporary knowledge, the idea that light's speed does not change is ridiculous. Holding onto these early twentieth century views, is choosing to be deliberately ignorant, similar to what the court of Rome did along with the clergy of the Church in the days of Galileo.

One major set of anomalies that can be attributed to the bent ruler of light is the farthest known galaxy, which is believed to have taken a half billion years to form and is claimed to be 13.2 billion light-years away from us. The "big bang" is supposed to have occurred 13.7 billion years ago. Do you see the problem with these numbers?

If the furthest galaxy is 13.2 billion light-years away, and if our Universe is expanding at a relatively consistent rate, then we must assume that the furthest galaxy is actually 26.4 billion light-years away—at a minimum. If it took 13.2 billion years for the light from that galaxy to get here, then the galaxy was in its particular, observed location 13.2 billion years ago. This means that if our Universe is expanding then the furthest galaxy would have continued to move away from us for 13.2 billion years from the location that we currently observe it at. Of course we understand that these quantities of years are so vast that our margin for error can be several scores of percentage points and no one will ever notice our errors. At this point, we simply cannot prove those numbers—they are utter speculation.

It is believed, and taught by many astrophysicists, that the furthest galaxy is much younger, about a half billion years younger, than the Universe's alleged birth date of 13.7 billion years ago. So what we are saying is that we are seeing light that is 13.2 billion years old that originated 13.2 billion light-years away; and further, we are saying that the particular galaxy was at that location 13.2 billion years ago. There are some glaring problems with this logic, but science seems to have perfected this sort of flawed logic over the years.

Another important point to scrutinize is science's twentieth century assumption and understanding of the Doppler Effect and the red-shift of light. Our knowledge of red-shift is based upon our assumption of light's unchanging speed, yet we use light's changed speed in the Doppler Effect to detect the red-shift. This alone causes many potential inaccuracies in our measurement of distant celestial bodies, and it puts considerable bends in the scientific ruler. Based upon Einstein's theories of the properties of light, the ruler of science is also bent by certain assumptions that space is expanded, contracted, or warped.

To minimize the years and give as much advantage to big bang theory as possible, the following chart assumes that Earth is at the far edge of the Universe and the most distant galaxy is at the other extent of the Universe. These positions are assumed to be in place 0.25 billion years after the big bang. These assumptions are made to align with scientific big bang consensus.

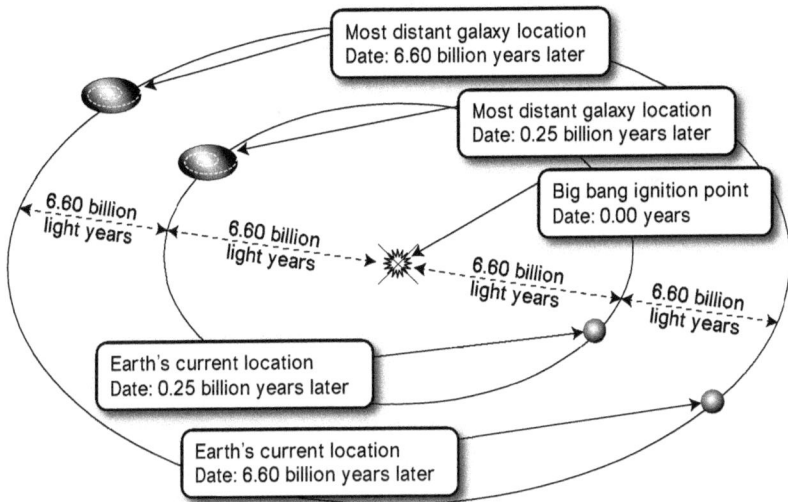

Figure 2 Universe Expansion

Slowing the radial-departure speeds down to the speed of light at the distances displayed on the chart, would put the actual distance, today, at 26.4 billion light-years, but the observable distance at 13.2 billion light-years. According to our current scientific understanding of relativity, there are many problems with the big bang, such as having to have to move at a minimum of 26 times the speed of light at a constant rate for 0.25 billion years. This anomaly is typically explained away by saying the "Laws of Physics" did not yet exist at the point of "big bang." It is true that the "Laws" did not exist because that was before humans existed to invent these arbitrary "laws." However, the functionality of physics that the "Laws", in futile effort, attempt to describe, likely was there before any bang could have occurred.

Chapter 4

Finding a Way

The human quest is to find a way to understand. So when we speak here of *Bending The Ruler*, we are not condemning the quest to understand; but rather, we are addressing the acceptance and deliberate attempts at permanently bending various scientific rulers that we partake in to cope with what we do not yet understand. It seems that the scientific mantra has become, "If it doesn't fit, then we will force it to fit!" That's like someone claiming that (G)god is on *their* side. Should we not, instead, say that *we* will abide by what is there, and adjust our descriptive mathematics to comply with what we actually see?

God Exists

A position held by many people is that Gods exists; and trying to debate with them about whether or not this is true is futile because their blind-faith is heavily set in their heart and mind. Real science-minded people "*know*" that there is no god... or do they?

Some people might believe that it would be good to write a science-oriented book and not discuss "God," but since this book is about *Bending The Ruler*, the topic cannot be ignored. This is especially so since a large portion of the world believes in a Creative God. At the very minimum, everyone must interact with people who believe in God, so it is important to properly educate them on what is believed to be found or known in scientific research.

Science-minded people often insist that there is no God, or better stated; there is no deliberate Creator. But, can a person with this attitude be considered as being truly science-minded? Can we say that we are open to truth when we insist that something does not exist without offering our proof of that perspective?

Just because billions of people believe in a Creator, does not mean that a Creator exists. However, it is also fair to say that just because many do not believe in a Creator, does not mean that a Creator does not exist.

If you are reading this and you consider yourself an objective person, but you *insist* that there is **not** a Creator, then, technically, you **cannot** be considered an objective person. For whatever your reason, you have a bias and a desire to explain away a Creator. Conversely, if, in your blind-faith, you insist that there *is* a Creator, and you refuse to receive evidence which on the surface appears to the contrary, then you have a bias that is based upon your blind-faith alone. Both of these views are identical in nature, and both are unacceptable scientific perspectives that do not allow for the discovery of truth.

We can harbor a belief of a Created, or of a creatorless expanse, but when that belief filters out real evidence that runs contrary to our individual belief systems, then we have doomed ourselves to great level of error.

If you insist that there is no possibility of a Creator, then you *cannot* consider yourself a true scientist, and thus, you have barred yourself from being able to see the truth. That statement does not say that there is a Creator; however, if anyone reads it as such, then their problems are far deeper than anticipated. If, in science, we insist that there is no deliberate Creator, then what else that we cannot see, do we insist does not exist?

Do You have a Religious Attitude?

When we insist that our way is *the* way it is then that we have a problem. We can try to convince others that our observations are accurate, but we cannot prove that an unobservable phenomenon **is not** or does not exist. When we scream or belligerently demand our way, we have adopted a religious attitude and we are no different than the Church leaders or government of Galileo's time. While the Church leaders funded Galileo's work, when the conclusion of the work was not to their liking, his work was dismissed, crushing the ruler of science.

It is very important to note that the Church is largely responsible for Galileo's findings through the financial support of much of his research.

Too many Christians who adhere to the fundamentals of their faith have adopted an ignorant stance that all science is wrong about the age of the Universe and evolution; they have thrown out the baby with the bathwater. However, science is no different, especially with regard to evolution.

Science is not something that is owned by someone, just as the Creator is not *owned* by someone. If you are a Christian and you believe that there is a Creator, then to further your understanding you should embrace *all* of the **observations** and **findings** of evolution—but not necessarily the **conclusions** of those observations and findings. Who knows, through those findings you may be able to actually *prove* the God that we speak of exists. If all of the scientists who are Christians, and the non-scientific Christians, truly believed in their God, then they would be leading science rather than fighting it. It is as if some people are afraid of what they will find, just like the Church Clergy was afraid in Galileo's day.

Additionally, we must ask: why is the science community so afraid of a Creator? What will it harm if there actually is a Creator? Not all scientists have a creatorless view, but it is a prominent belief in the science world. It is odd that some people have an agenda to bar the belief in a Creator from the minds of others. This behavior is identical to *stereotypical* Creationist Christian behavior that is condemned by many evolution and big bang supporters, and it is identical to the attitude that the Clergy had against Galileo.

To call yourself a scientist you must be willing to consider <u>all</u> ideas. You do not have to embrace or believe all ideas, but to discount them based upon your own blind-faith biases, is to be deliberately ignorant; and thus, it is bending the scientific ruler and tying it tightly into knots. Claiming that there is not a Creator without having any proof is as full of blind-faith as Christians who claim (based solely upon the Bible) that there is a Creator. If there is a Creator, then there is likely to be evidence somewhere of a Creator, so then, why the fear of discussion of one perspective or another? If there is a Creator, then we should be excited because there is someone watching over us that is holding all things together. If there is not a creator, then we have no hell to worry about, so why the fight from either side of the discussion?

Casting away blind-faith and dumping it into the abyss is a win-win situation for everyone! If creatorless scientists dump their blind-faith that there is no creator and if they are correct, then they have nothing to lose in opening up their minds. If Creationist Christians are correct about a Creator, then they have nothing to lose by also embracing *observations* made by Darwin and other scientists. However, Darwin's *conclusions* based upon the evidences found by science, are an entirely different issue than embracing his *observations*. For instance, Christians must realize that it is not Darwin's observations that are in question, it is his **conclusions** that are the questionable issue—Creationist Christians are actually disagreeing with Darwin's **conclusions**. Only those who are afraid of the truth will bend the ruler enough to ignore what is obvious. This goes for Creationist Christians, as well as for creatorless scientists.

Laws?

Why is there such a chasm between Christianity and science with regard to big bang and evolution, versus Creation? The chasm is because of *laws*.

You might wonder why *laws* are the problem, but we do not need to look far to see the devastating effect of trying to impose our will on others; recall World War II, or the Inquisitions.

The scientific "laws" are *observations*—not laws. And believing that these "laws" govern what we see around us, affects our scientific ability to see and find the truth about the way things really occur. Why do we believe that the Universe and atoms must follow *our laws*? Our laws are only flawed and very short-sighted descriptions of how we *believe* the things that we observe function.

We have gone so far as to place a "speed limit" on the speed of light. If asked, many of us understand that these terms are there to convey our observations, but we do not respond to these "laws" as descriptions; rather, many respond to them as concrete inarguable factual laws that cannot be violated.

In reality, the jury is still out on most of what we call the "laws of physics." These laws should be referred to as "descriptions" of physics. Incidentally, the etymology of the word *law* according to Webster is "*lie*." Believing that what we *think* we *know* about physics are actual *laws*, makes Webster's etymology quite appropriate.

Our Math is Finite

Our math is finite to our experience, and it is designed to measure what we see or can observe. Our experience is as finite as we allow in our minds. We invent math to explain to ourselves what we have observed. We do not yet know the various curves that must be built-in in order to adjust for size errors, or size-caused modifications in our calculations, such as the size, mass, and distance of a star.

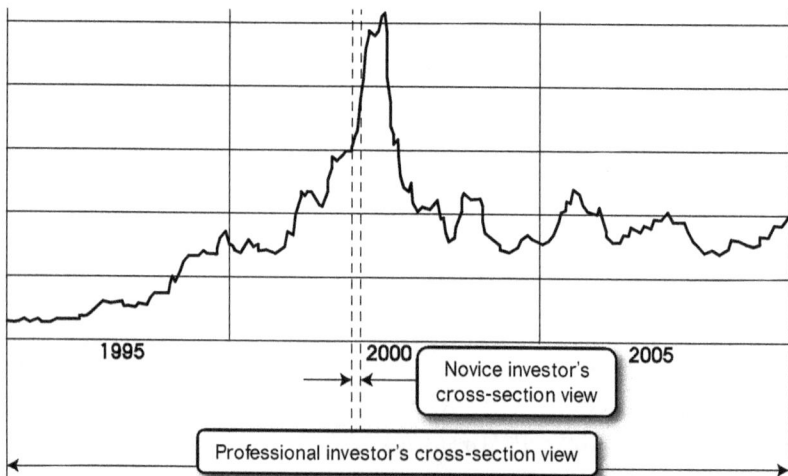

Figure 3 Stock Chart

If successful Wall Street stock traders functioned like scientists do, then they would take a tiny slice of a stock chart and base all of their timing decisions upon that tiny slice of the chart, and in doing so they would not be successful for long (Figure 3 Stock Chart, Page 70.) This is what happens during a stock bubble: Inexperienced stock traders and experienced stock traders alike, fail to look at the bigger picture, and then they extrapolate their entire broad view from the tiny finite slice from the graph of the current trading period. Doing so causes them to assume that the stock will continue to increase at the same rate for an undetermined period of time. Later, when the bubble bursts, they run home crying that they were cheated, but, in truth, they only cheated themselves because they failed to understand the *entire* graph.

Just as science did throughout its history; today it continues to use the exact method that inexperienced and foolish stock traders do. At some point, in stock trading, there will always be a normalizing of the rate of change and the stock will drop as the bubble bursts. The only problem is that with science there is no bubble to burst, except for within the heads of some scientists, so the errors are allowed to remain hidden by the bent and contorted ruler of science. They are hidden until a scientific heretic with a bold soul dares to defy the religion of science and prove it wrong. We do this same thing with our physics math; we take the tiny finite sliver of what we observe and demand that *everything* fit that model or law.

Because we can assemble an eloquent equation that appears to perfectly describe one specific aspect of the observable Universe, we make the grievous error of assuming that the same math applies to all levels of physics. This ignorant approach is the cause of our warping of space and time in order to make our numbers work out the way that *we* want them to, and then the problem gets worse.

Finite extrapolation is a behavioral issue. If we use finite extrapolation with time and space, then we will do it wherever we need to in order to "prove" our theories. This behavior is being inadvertently taught to each new generation of scientists that enters the education system, starting from a very early age. This occurs because we fail to teach the teachers and the students to ask "why?" in an effort to better understand anything. Instead we just tell them, "This is how it is, now accept it!"

Our inaccurate use of finite mathematical formulas is able to hide behind its usefulness with regard to the finite portion of physics that the formulas initially were formed to describe. Just because Newton derived a very useful formula to describe gravity, does not mean that it applies to everything. This is just like the previous chart's novice perspective where it only applies to a particular and small portion of the chart. This finite slice should not be extrapolated across the entire chart's spectrum because it will likely be increasingly wrong as the distance from the event-moment widens.

Our mathematical extrapolations can reveal a great deal, but we must realize that they are likely flawed, thus hiding much of what we desire to discover.

Practicality

In the beginning of this book I mentioned *practical application.* Humanity, and especially science, believes in practicality, but we fail to see that the Universe may share that same property. Instead we chose to bend time and space. Since "chaos" cannot be specifically demonstrated, and to keep things easily understood, we will use the definition of *unpredictable.* The word **unpredictable** means, "not able to predict." This causes need to understand the word **pre-dict**; *pre* being "*front*" or "*before,*" and **dict** or **diction** meaning to "*say.*" Therefore, the term *chaos* is a short-sighted self-notification that we don't really understand what we're talking about and we have not yet found a means for understanding it. It's okay if we use the term "chaos" to describe our not knowing something, but if we insist that something is unknowable, then we are making an error.

Anything that moves as a reaction is predictable, however, when outside intelligent-intervention occurs, then we lose our ability to scientifically predict. We tend to ignore intelligent intervention, yet we control where an atom will be at any particular moment when we move our own bodies. From a scientific point of view this could be considered chaos because the movement is based upon the thoughts of the possessor of the specific atom. Unless we can predict what a person's thoughts are, we cannot predict where a particular atom will be. This could be considered chaotic, but that is not a fair assessment because that atom is being masterfully controlled by the possessor of the atom. Science does not account for this.

Science, religion, and philosophy are truly one in the same. They are all an effort to understand ourselves and our environment. When we accept things like "chaos" we have forfeited our ability to know something. For instance, expecting to understand detailed specifics about the weather is a foolish endeavor because the butterfly-effect is, at least to a certain extent, very real. Small and unrealized effects of slight variations in temperature and moisture-content in land mass, obviously affect weather in subtle ways too difficult to calculate adequately. However, we can make macro predictions based upon history, broad assumptions, and obvious patterns of motion, which is why weather prediction accuracy varies greatly from region to region and is basically the toss of a coin. It's seldom that the weather forecast for a given day remains constant as the given day approaches. While we are able to speculate on weather, the unending variables are far too numerous for us to gather and process quickly enough to achieve little better than a guess of chance after looking at the basic flow of the coming weather systems. Often a flip of the coin is as accurate as the extended weather forecasts, and in many cases the flip of a coins is more accurate than the weather forecast.

The only true chaos is the action of living beings, and even that cannot be called chaos because we make choices, and then those choices control the elements within us as well as those that surround us. Our ability to understand the predictability of any aspect of the Universe is limited to the accuracy of the rulers that we are using. When we bend the ruler to fit, then we lose all practical ability to measure anything properly and accurately.

The practical accuracy needs of our measurements depend upon what we are measuring. Machinists typically measure within thousandths or ten thousandths of an inch. Carpenters measure within eighths or sixteenths of an inch. Excavation crews measure within a few inches, and astrophysicists measure within plus or minus thousands, millions, and even billions of light-years. Grade school children are far more accurate when it comes to measurement than astrophysicists are. And grade school children measure using only the shiny new rulers the children get at the beginning of each school year. Where did we go wrong? A child is more accurate with a ninety-nine cent ruler, than many adult scientists are with billions of dollars of scientific equipment! Odd, isn't it? The people that we perceive as the most "sophisticated and educated" are the most inaccurate.

We must stay practical and insist upon finding math **and** words that adequately describe our Universe, rather than trying to make the Universe fit *our* math and *our* understanding of words. It could be argued that this is what many of the physicists do, that they try to make formulas that describe the Universe. This is true, but there is one glaring problem: Some of the indices used in our formulas are questionable and are likely flawed, as is made evident by the fact that they do not explain the anomalies often incurred. This potentially makes our equations grossly misleading.

Is Light Subject to Motion?

Light's properties is one particular area where science may be grossly misled. Here is a good question to think on: If you take a laser beam and shine it into space and then wildly swing it back and forth, does the light beam get cast in a sideways motion? In other words, if you move the light beam abruptly to a sideways position, can the light beam swing, or be thrown from momentum, beyond its hypothetical limit? This is as opposed to the theoretical straight line you get when you point the light beam steady in a specific direction (Figure 4 Light Mass Over-Bend, Page 75.)

As seen in the overhead view in the following depiction, if light is affected by gravity, then this should happen in a curve. This does not depict the plotting curve that will occur naturally when tracking the consistent change of distance versus the progressive acceleration change in sideways motion. The light beam length would be very long and the amount of curve would be dependent upon the sideways velocity. The following diagram assumes an instantaneous halt in sideways motion. Also, would the effect increase as the distance from the light source increased?

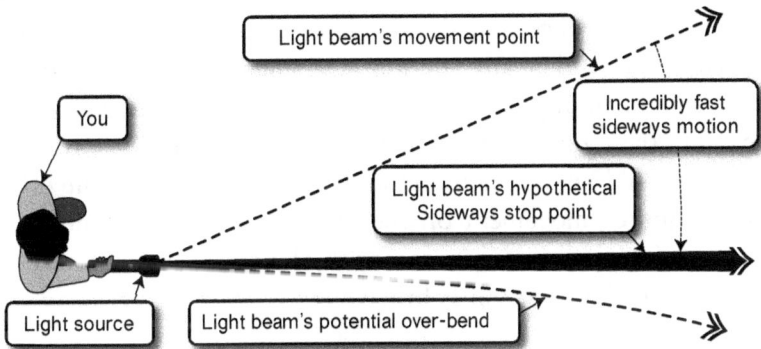

Figure 4 Light Mass Over-Bend

Light's peculiar and seemingly massless nature leads us to believe that light is constant, yet we believe that light cannot escape a "black hole", or better stated, a dark star.

If a black hole or dark star does have an "event horizon", then at some point, light has to be slowed down. Based upon conventional theory, if you were not crushed by the gravitational forces of a dark star, *for you* (being inside of the event horizon) time would be normal; whereas for the outside observer, you would be standing still *if* you could be seen by the observer. But, since the light theoretically cannot escape the event horizon, it theoretically cannot be observed.

All of the theories about light, such as "event horizon", are built upon the debilitating assumption that light is constant; yet, it seems highly unlikely that light is constant. However, it's fair to say that light is the most constant thing that we know of today.

Light's unique lightweight or apparent massless nature, in a practical thought process, causes illusions in our minds that make light appear to be something that it is not. For some reason, we have a difficult time grasping the actual speed of light and its apparent masslessness.

From a practical perspective, if light is nearly massless and nearly unaffected by gravity or other substance, then it would be expected that, due to the instantaneous nature of light, the light emitted is mostly unaffected by any motion that we can effectively produce and detect. The photons are emitted at a speed so great that any motion is negligible in relation to the light's moment of photon release. Since a photon is released in an instant, then only the motion during the release portion of motion could be considered in an equation calculating the effect of the source movement on the emitted light. That moment is essentially zero in comparison with the speed of light and the apparent masslessness of light.

The hypothesis behind this thought is: if a source of light is traveling at the speed of light and is moving perpendicular to the direction of the emission of the beam, then that light would travel approximately at a 45 degree angle to where the photon was released from. See the following illustration (Figure 5 Sideways Motion of Light, Page 76.)

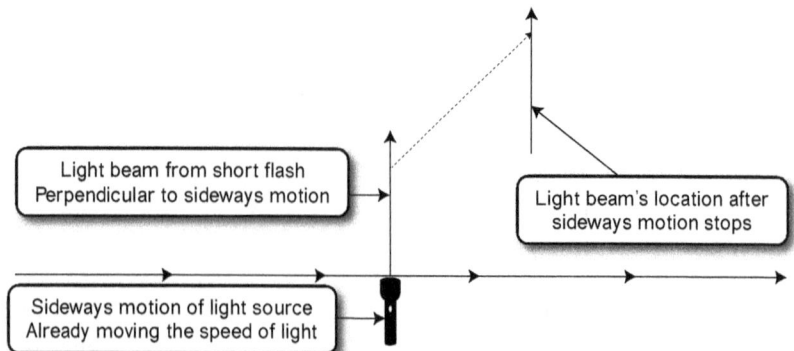

Figure 5 Sideways Motion of Light

Light beam from short flash Perpendicular to sideways motion

Light beam's location after sideways motion stops

Sideways motion of light source Already moving the speed of light

Because we currently cannot travel at such speeds, we have no way of truly knowing this. Light appears not to be subject to motion because it is launched instantaneously, and it is massless, or close to it. It is likely that either light's speed, its apparent lack of mass, or both, make light appear as if it is constant. If something was able to move fast enough, then it could overcome the lightness or the assumed masslessness of light, which would likely affect the speed of the light and the light itself. This practical view is simple, but allowing light to be variable can help scientific understanding a great deal. If our Universe is moving at the rates that are assumed, based upon the big bang theory and "the furthest known galaxy", then we are likely already moving at or near the speed of light, if not in excess of it. Given our big bang viewpoint, believing that we are not moving at the speed of light is highly contradictory.

In reality, it is unlikely that the speed of light is constant. Light only appears to be constant because of its instantaneous nature, and also appears so due to our inability to make observations at velocities great enough to cause discernable differences. Einstein's theories indicate that from an observer's perspective, light is always constant. This is because an observer cannot alter their speed enough in order to affect light enough to make any measurable assessment of the effect of the movement during emission of light from the observer's source movement. Claiming light's speed to be constant is most likely incorrect, yet light is the most stable thing that we know; so, for all practical purposes, light is constant in relation to our ability to measure any usefully discernable variation.

It is only when we use light to assume tremendous distances and great lengths of time that its subjective nature to motion begins to be revealed. Instead of embracing that light is variable, we, in our efforts to find a way to understand, have chosen to bend the most important ruler that we have—we have made light our god of science!

Chapter 5

Breathing Life into the Universe

To diverge a bit from the celestial and quantum realms, since we humans are the ones who are doing all of the research, a discussion of *us* is in order. The very fact that humans live and breathe is an amazing function in itself. These human functions give good cause, in many minds, to believe that there is a deliberate Creator.

Just how did it all come to be? We have a tendency to point to Darwin as being the father of evolution, and while he was the first to lay down a clever argument for evolution in a book, creatorless creation is nothing new. For a very long time, varying beliefs on this topic have ranged fully throughout the entire belief spectrum. The debates of a god, gods, and existence ending at death, versus a life hereafter, have been major points of contention throughout recorded history. This same problem continues today, and it will likely continue in the future.

Take for instance the godless view of the world: Many of the proponents of a godless world are on a mission to discredit the Bible. However, through its historical content, the Bible is being proven more accurate every day; yet, many people hold fast to the belief that the Bible is a bunch of made up stories.

We can debate about the more fantastic aspects of the Bible, but trying to discredit it due to some perceived discrepancies puts us in a dangerous position of being wrong. For those who choose to open their eyes enough to research the topic without bias, many of the historical accounts in the Bible have alternate reinforcing sources. Yet, the many and glaring inconsistencies in the evolution and big bang theories are overlooked and utterly ignored while big bang and evolution are called "fact" by many scientists and are sanctioned as such by governments.

The vast historical value of the Bible does not mean that it is accurate on *all* accounts, nor does this mean that there is absolutely a Creator. The evidence of whether or not there is a Creator is all around us. What we choose to attribute our surroundings to is a personal choice. But, no matter what we attribute our surroundings to, in the end, only those who are correct in their choice will be correct, and the rest of the people will not be correct because they failed to embrace truth. As for now, we will have to wait and see who is correct until the overwhelming evidence comes to light in the minds of the people. Let us not be people who vehemently deny that a Creator exists, but then go on to insist that "life" came to Earth from aliens from another planet or from some space debris that struck the Earth. While this could be true, we, then, still have to discern how the alien life form in its place of origin came to be to begin with.

Because we are here and we must all coexist, it does not matter if you think that a deliberate Creator did it all, or if you think we were delivered here by aliens, or if you think that we have evolved from some sort of protoplasm. Our choice of which of these perspectives we have chosen to believe, and *how* we choose to believe each of them, greatly affects our ability to be scientifically objective. Each perspective has its own agenda. Our problem is in discerning which is correct. It is possible that we are all wrong, but I can say with a great amount of confidence and certainty that whoever has chosen to seek the origins issue accurately with an *open mind*, stands a far better chance of revealing the mysteries of the Universe while being credited for their own tremendous contributions to science! Those who get it right and are the first to propose the accurate information may even receive the coveted Noble Prize. Yet take heed when seeking such prizes, because many such awards have been greatly devalued over time due to frivolous awarding for erred theories and pumped up notoriety. Some awards have even come to a point where they might *not* be given to someone for *accurate* and *provable* discoveries because the discoveries do not fit with the consensus of the contemporary scientific *belief* systems.

Is all Matter Created from Nothing?

To shortcut science, all of what we call "*matter*" is *created* from nothing, or it is "*Created*" from nothing with a capital "C" (In saying "from nothing," I am referring to it in a purely detectable physics sense.) I, for one, am glad that we have been trying to seek how the Universe has occurred, regardless of whom or what caused it. To me, particle physics is the most interesting of the sciences, even though, in the end, we will likely come to the conclusion that it is all made from nothing. But how? We should never stop trying to learn how everything and everyone truly came to be!

In saying, "how everything and everyone truly came to be", I do not mean Creation versus the big bang and evolution. Rather, I mean, *regardless* of how it all came to be or who did it all, we should, without question, try to understand how it all came to be. It is the ultimate scientific quest, and it is incredibly interesting! It's also practical and useful to us because we can use this understanding to improve life in general. And we can use it to learn the deep secrets of the wonders around us.

To lack the desire to understand the details of our origins would be like not investigating the pyramids or some ancient culture just because we know that the pyramids were built with really big blocks of a stone-like substance a long time ago. Exactly what did the Egyptians do to entertain themselves; did *they* study an even more ancient culture that we know nothing of today? It's likely that they asked questions similar to the questions that we ask, such as "Why do the heavenly bodies move as they do?" From a Biblical perspective, the Egyptians may have known their origin in a very real and connected manner, which would be dependent upon their belief set, because some of them may have, theoretically, actually known Noah. Regardless, we do know that the ancient Egyptians had a certain level of obsession with death, and, also, an obsession with some sort of after-life and the celestial bodies, making it obvious that they did contemplate these things to some extent.

We are Sentient Beings

Sentient beings are physical beings, meaning that we are able to perceive or feel—we are conscious beings. All of our research is done to satisfy our curiosity about our surroundings and about what we are constructed of. Our chemical bodies touch, taste, smell, hear, and see. It is the things that we perceive with these sentient senses that we want to understand.

For many of us, there is a quest to step beyond our physical sentient nature and delve into the spirit realm. These lines understandably blur for many people when discussing quantum physics and metaphysics. This is because quantum physics does not pursue the intangible, but metaphysics does; and these two areas of discussion often play at different corners of the same sandbox.

If we actually all have "spirits," metaphysics allows for those thoughts and debates to be in the discussion, whereas quantum physics does not involve the spirit, and typically rejects the possibility altogether. To describe it in non-philosophical and non-religious terms, metaphysics is sort of a combination of quantum physics and psychology. Ignoring these aspects of discussion in science has become the norm over the years. But, we should not disregard the possibilities of another realm.

It's curious that we can bend the ruler of time and space in order to force light to be constant; and through this we insist that there are more than the four basic dimensions and that there are parallel universes that reside where we reside, within the same space but in a different dimension; yet, we will reject a *possibility* of a Creator and a place or dimension commonly called Heaven. This makes no sense; it is similar to insisting that life on Earth came from aliens, while insisting that there is no god. Where did the aliens come from? How did they get there?

We end up chasing a carrot on a string when we choose to place our origin in another place. If life came from Mars, then where did the life on Mars originate? This manner of thinking is infinite in nature because you could keep passing the buck to the next celestial body and its satellite bodies. However, then you will keep having to explain the same ridiculous argument—infinitely; thus, avoiding the task of actually answering the question, "How did we come to be?" At some point, we need to accept the logic that we are constantly setting up barriers when we chase the origins-carrot-on-a-string in this way, and, for whatever the reason, it seems natural in us to do so.

The "life came from aliens" argument is similar to a multiverse as opposed to a single universe. Proclaiming that a multiverse exists is just setting up another wall against infinity, and doing so will lead to an obvious next level to set up another wall between us and infinity. We could call that next imaginary level a multitude-verse, or a multi-multiverse, or something to that nature.

Our sentient senses struggle to be able to fathom a few fundamental ideas such as: *infinity, expanse, firmament,* and *time.* But infinity is the primary of these. If you cannot wrap your mind around infinity, then you cannot wrap your mind around **void,** an **absolute,** or the idea of **eternal.** The big bang theory exhibits this same infinity-of-walls issue because the big bang is believed to be cyclical. Infinity is more of a metaphysical point of thought than it is quantum physical in nature. Infinity exceeds our sentient nature and therefore we struggle with it in science. We, as sentient beings, are always trying to quantify everything by measuring it and placing it into neat little boxes with dimensions that we can comprehend.

How many dimensions are there? In consensus, four at last count, unless we want to count motion as a fifth dimension. The other dimensions are length, width, height, and *time* or how long something has been there.

The illusive fourth dimension of *time* does not compute with our sentient nature, so we bent it and we bent it good! ***Time...*** Can we touch it? Can we feel it? Can we see it? Can we detect time with our sentient nature? We claim time to be a dimension. We even claim that we can measure it. Does time have mass? Does motion have mass?

Where, What, When, Why, Who?

Where, what, when, why, who—did you ever consider these words in a group? Take a few moments to contemplate them. Take time to understand all of these words because there is a great deal of understanding bound up within them in regard to our non-sentient nature, as well as our sentient nature.

Where is the void of space. **What** is the firmament and that which is held within it. **When** is the moment in time that anything is there, which is where science ends its case. Science would add to this list of words **how** rather than **why** because "**why**" gets more philosophical and indicates reason or purpose. The **Who** aspect has many religious implications; philosophy does not necessarily address the **who** aspect, but religion does address "***who***" by utilizing a (G/g)od. It just might be that science, philosophy, and religion are an inseparable whole that we refuse to see as one.

In an attempt to explain what we can't explain, we bend the ruler with each of the questions *where, what, when, why, how,* and *who* so that we can pretend that we see our way clear to understanding. If we choose to look, we can often observe ourselves inventing far reaching fantasy ideas in order to explain what we do not yet understand. If we're too lazy to make things up, then we just say "God did it all" or "it all evolved" and we ignore the many unanswered questions that we leave behind in our trail of self-deceit.

Chapter 6

Is Our " Scientific Method" All Inclusive?

In the science world we believe that we are open-minded and objective, but are we really? Is our "scientific method" all inclusive? The simple answer: No!

Science should not get a pass in this world when being dishonest, because it is no different than any other industry. People steal each other's ideas and take the credit for themselves; but even though there are many selfless people in the field of science, arrogance does prevail, and some scientists will even lie, cheat, or steal in order to be recognized.

The scientific method, as defined, is fantasy. It is a fantasy because it fails to include the deliberate ignorance of many facts; and it also fails to include the selfish nature and financial- and agenda-driven aspects of science. The scientific-method ruler is damaged just as badly as the ruler-of-light is. This makes perfect sense because a tremendous amount of our science is based upon our belief in the constant speed of light, and in other potentially error-prone theories.

The scientific method excludes several critical factors. As mentioned a moment ago, theft and arrogance are not included in the calculations used in our scientific method. In addition to that problem, actual accurate data is often missing in the scientific method when we deliberately remove certain data due to the fact that we don't grasp it. When we do these things, then our "laws" of the scientific method quickly break down, but we ignore that to rationalize the things that do not make sense to our minds, like saying that that laws of physics break down when approaching singularity.

Potential Gross Error

The different views of *mass* in *special relativity* and *general relativity* are often disassociated; wires are crossed and the lines are blurred. Our bent ruler of the properties of light offers us potential for tremendous error. Since the margin for error is so great when calculating vast distances in space, our errors will increase proportionately with the distance being measured. And depending upon the type of error in our math, and the particular equation being used, the errors can be exponential.

If Einstein's "*constant*" is incorrect, then any calculations made using that constant will have exponential error. This potential gross error will be undetectably small with regard to anything that we actually physically experience up-close-and-personal. However, when calculating distances as vast as our Universe, these exponential differences can become incredibly large. If our assessment of light is inaccurate, then our margin of error on vast intergalactic distances could well exceed 75 percent.

Some theories of the big bang have bent the ruler beyond recognition and have made the big bang expand at a speed greater than the speed of light, which is believed by many scientists to be impossible; thus the "*the laws of physics break down near singularity*" belief is born. Big bang theory speeds bang expansion to thousands, millions, billions, or even trillions of times greater than the speed of light. Even so, if the Universe expanded at only twice the speed of light (if we are at the center or start point of the big bang) and the light that is reaching us now was emitted 13.2 billion years ago, then this 13.2 billion year old light would have passed our stable central point of location 13.2 billion years ago.

If the Universe expanded at twice the speed of light, while moving us and the furthest galaxy away from each other in opposite directions, then it would have taken 6.6 billion years for the furthest galaxy to move 13.2 billion light-years away from us. If this were the case, then that means based upon fundamental mathematics, that the light and radiation emitted 13.2 billion years ago, when the furthest galaxy was allegedly formed, is far beyond us now.

Since the furthest galaxy is supposedly 13.2 billion light-years away from us, based upon our interpretation of current light and radiation observations, the actual location would be 26.4 billion light-years away since it is moving at a rate double that of light. Therefore, the *current* emitted light will not reach our current location for 26.4 billion years. This does not include the potential implications if light has any mass whatsoever.

If the Universe did expand at twice the speed of light, then the latest that the furthest galaxy could be 13.2 billion light-years away from us, in its current location, is 6.6 billion years. If our distance from that galaxy is 13.2 billion light-years away, this means that the light and radiation that was emitted 6.6 billion years ago would have been emitted from 13.2 billion light-years away. This would put the returning light from the emitted location in space and time only halfway back to us. This means we would not yet be able to see it for another 6.6 billion years. Having moved at our assumed consistent rate of twice the speed of light, the closest possible observation point would currently appear to be 8.8 billion light-years away.

Does any of this confuse you? I hope it does. If you're not confused by now, you must be lying to yourself somewhere in the equations in order to justify the big bang as it was presented at the beginning of the twenty-first century and before that time. We cannot have it all ways: we cannot lie to ourselves and still have the math work accurately. The math breaks down immediately upon trying to work this out because there are errors in the logic. The only way that the light can be here now from an object that is 13.2 billion light-years away is if that object was in the location that we currently see it in, at a time of 13.2 billion years ago.

For my part, please disregard the discrepancies in some numbers seen in the diagram (Figure 6 Big Bang Expansion Timeline the Universe, Page 90.) I am using the key age and distance numbers given to the world by the scientific community at the time this book was compiled.

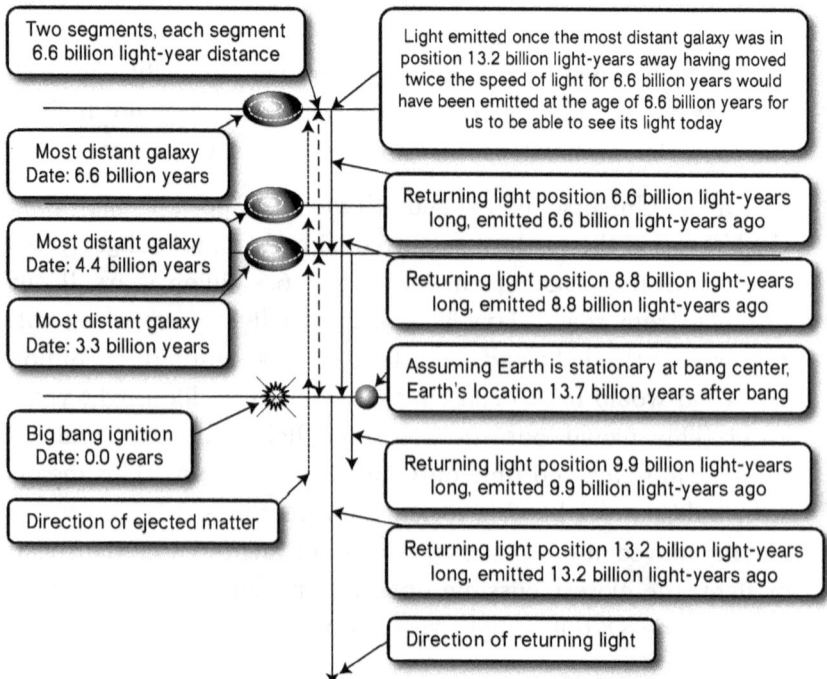

Two segments, each segment 6.6 billion light-year distance

Light emitted once the most distant galaxy was in position 13.2 billion light-years away having moved twice the speed of light for 6.6 billion years would have been emitted at the age of 6.6 billion years for us to be able to see its light today

Most distant galaxy Date: 6.6 billion years

Returning light position 6.6 billion light-years long, emitted 6.6 billion light-years ago

Most distant galaxy Date: 4.4 billion years

Returning light position 8.8 billion light-years long, emitted 8.8 billion light-years ago

Most distant galaxy Date: 3.3 billion years

Assuming Earth is stationary at bang center, Earth's location 13.7 billion years after bang

Big bang ignition Date: 0.0 years

Returning light position 9.9 billion light-years long, emitted 9.9 billion light-years ago

Direction of ejected matter

Returning light position 13.2 billion light-years long, emitted 13.2 billion light-years ago

Direction of returning light

Figure 6 Big Bang Expansion Timeline the Universe

How do we Measure Time Travel?

Our concept of time travel is a bit of a misnomer. What is time travel? How do we measure it? Einstein's formula does not allow time travel, but quantum physics has bent the ruler a little further to allow for it to happen—in theory. This idea of traveling faster than the speed of light somehow allowing us to have the ability to move through time has a few glaring over-sights.

The first problem in navigating time by exceeding the speed of light is—moving faster than the speed of light in reference to what... the lone individual? Can moving backwards at the speed of light move us backwards through time? Backwards, in reference to what... forward? With such logic we could turn our body to face a different direction and we would suddenly be time traveling.

Edwin Hubble's observation states that all objects observed in space are found to have a Doppler shift observable-relative-velocity to Earth and each other. If the Universe is actually expanding as indicated by Hubble's Law, then we are probably already moving at the speed of light or in excess of it because the residual expansion of the big bang would still be moving at tremendous speeds. Given the last statement we could hypothesize that the Universe's movement is what moves us forward through time. If it is reversed in direction then everything would go back in time instead of forward. The idea of velocity allowing time travel is from the perspective of time travel enthusiasts.

When we take existing "laws" that may be flawed and add to them more laws that may be flawed, such as Hubble's law (which is based upon the previous laws), our adding together of these manmade laws of physics hinders our forward progress. And often the inherent errors in the flawed laws compound each other. Building bent premise upon bent premise will not get us to where we want to go. When we build laws upon a foundation of error, then our new laws will most certainly be flawed and inherently carry with them the errors of that which they were built upon.

We can't truly even say that the clocks we use are accurate given the math used in physics. Bent indices make for bent results. We believe atomic clocks and atomic half-life to be among the most accurate means of time-keeping, and it very well may be. However, if satellite clocks lose time because of gravity differential in orbit, or because of their speed relative to Earth, then are they really that *accurate*?

What our atomic clock system does for us is to synchronize *us*, but the time keeping is only as accurate as its basis. If the clocks we use were "accurate," then adjustments would not ever need to be made. Did you ever notice that we *do not* have to adjust the Earth and that we *do* have to adjust our clocks? Our day-to-day terms of *time* and *day* are rooted based upon the Earth and its rotation. In regard to the idea of time, the Earth is chief and *it* dictates how we adjust our time.

We see the effects of gravity on our manmade clocks, but we fail to apply that same knowledge to our assessment of the age and size of the Universe. We also fail to apply this knowledge to the speed and properties of light. This is the root of time travel—no bent ruler...then no time travel!

While some of what we see in the gravitational and/or speed effects on satellite clocks can be explained by Einstein's theories, it does not mean that we are bending time or space when we move through space.

The Earth's spin is far more accurate and reliable than our atomic clocks will ever be. After all, isn't this what we are measuring to begin with, a single rotation of the Earth—a day? But in our human nature, we even have debate regarding how that is to be done.

Bending Time and Space

Recalling the stories about the *War of the Worlds* radio show fiasco in the late 1930's, we can see that the mindset of people during the early twentieth century was quite willing to accept fantasy based information contained in the radio show and believe it to be true. The bending of time and space was ripe for the picking at that time. It seems peculiar that civilizations are supposed to be getting smarter as a result of natural selection's evolution, yet, a couple of decades after Einstein's theory of general relativity, there were people who believed we were being attacked by aliens because they believed what was being broadcast in the fictitious radio show.

Both quantum physics and astrophysics are somewhat arbitrary forms of science because it is mostly guesswork at this point; much of what is being proposed in those fields is better fit for the sci-fi movie screen, than it is for the science lab. What we see, versus our conclusion about what we see, are two entirely different topics. This needs to be recognized *sooner* rather than *later* by the scientific community and by the general populous.

We create arbitrary mathematical rules, and then, with those rules, we claim our speed bends time and space. Exactly how egocentric is the human race? Is *our* speed really going to slow time? We want to *bend* time and space, yet we always refer to it in a constant linear manner when we try to measure the size of the Universe.

We should not oppose attempts to speculate the age of the Universe, but we should oppose any bent mathematical reasoning used in doing so. Some people of the Church have had in the past, and some still have today, the fearful point of view that we should not try to search out these seemingly unexplainable questions in order to find out the way things actually work. Science is not any different because if you deviate from popular scientific consensus the science community will excommunicate or disregard you. We act as if, somehow, God is going to zap everyone for being curious. We *should*, and we *must*, look into how it all came to be! Whether we have inborn curiosity from a Creator, or we evolved from protoplasm, we must remain curious! Anyone who tries to inhibit alternate ideas from being revealed should not ever be trusted. However, we must not confuse this with a necessity to demand evidence from someone who is proposing far reaching theories that do not properly align with evidence and physical reality.

A few main levels of certainty have been proposed over the years: *hypothesis*, *theory*, and *law*. We believe *law* to be absolute, but it is not. Like in the earlier example (Figure 3 Stock Chart, Page 70.), we might only be looking at a small portion of the graph and not realize it. There is only one truth, and it is *our* quest to find that truth. Until we find truth, we cannot say that something is "*law*." The moment we believe something to be "*law*", we have blinded ourselves with the constraints of that particular bent ruler of law.

Einstein bent time and space because of the belief that "light" is constant no matter how fast or what direction you are traveling relative to yourself. This sounds like a criticism of Einstein's theories, but it is not; it is a criticism of our willingness to except it as "*law*", or even as an irrefutable "*theory*", and the assumption that what a man said is the absolute truth, especially when even Einstein himself was not certain. To test Einstein's personal amount of certainty, we would had to have challenged his hypothesis with the threats of his death, and I doubt that anyone seriously did so. Would Einstein have demanded his way in the presence of his G/god if he was shown something else that made more sense? Likely not.

The way we have chosen to understand the "physics" of the tangible world is to embrace Einstein's theory. His theory bent time because, at that time, *he* could not see how to make it all fit together. He didn't have some of the information that has been made available since then. Einstein may have figured out his coveted unified theory if he had released light from *his* constraints and allowed it to change as needed.

Since the early twentieth century's proclamation of general relativity, we have seen a great deal of additional information. Einstein would likely be first on board to do a substantial bit of re-analysis of his own theory in order to understand the rest of the story.

Don't Confuse Tangible with the Intangible

Our concrete experience in life causes problems with regard to our perception of reality. In days long gone, we believed that the ancients may not have grasped the idea of air and gases because they had no way of detecting such seemingly intangible matter. But for us, since then, we have found many ways of detecting gases, and we understand that they are very real.

We are no different today with regard to light, time, gravity, and matter. Because we cannot see or experience certain aspects of them, we do not believe them to *be* (to exist), so we invent obscure theories in attempt to explain away our errors.

We need to decide what "*tangible*" and "*intangible*" are, which brings us back to the definition problem. **Tangible** means "*touch*," like with a **tangent** where one line *touches* another line. If something is tangible then we can touch it, or it can touch us, such as light illuminating us. We can feel the heat from the spectral radiation from light because as it radiates towards us, it ultimately touches us. Regardless of our chosen perspective, we do *feel* or sense light. Can the same be said for gravity? Does gravity touch us? Can we call it tangible?

Anything affected by gravity is tangible, but the question remains: is "*time*" affected by gravity? This is where my point of contention with science's blind acceptance of Einstein's theories is derived from. Just because someone said something that seemed to make sense at the time, does not make it so. We can use incorrect information successfully, provided that its errors are within our tolerance needs.

We must not confuse the gravity or electromagnetic spectrum with time. Time is something that is not tangible, and time cannot be altered, but the measurement of it can.

Dangers in Absolutist Science

Any absolutist science is subject to error; and our view of the term "*law*" is inappropriate. The inaccurate "**laws**" of physics are similar to a bad law that a government imposes on its people. A bad law should be rebelled against and reversed so as not to oppress the people. Absolutist science does exactly that, it oppresses the people; and it's not just the science community people; but rather it is any of the people who dare to speak against the law who will be oppressed by the Pharisees of science.

Our religious style, absolutist-belief-method, is truly the only damaging thing in our world, and it is what all wars were founded on; "It's my way or the highway!" This serves no one, and it is an attempt to assume authority. The only true authority can only be given by truth; and the closer to the truth you are, then the more authority you have. This applies to all of life, including science. Though, merely claiming you have truth does not make your claim true. Claiming that you have truth when you do not makes you a liar, or at minimum it makes you wrong.

Acting as an authority does not make one an authority. Science is full of pretenders who have been "educated" about all of the "laws of physics." For that part, they are the authorities on the "laws" of physics, but this does not make them correct. It is only in our own minds, and the minds of our accomplices, that our "laws" are correct.

Let's break through this sort of arrogance and allow truth to come into the mix. There are young students (scientists to be) working at the local restaurant who have far better and more accurate ideas about physics, than do the best of the best of the learned physicists and biologists. Youth has a great advantage in regard to absolutist ideology, that is until they attend school and are told "how it is." Once they are indoctrinated with the "laws" of physics and even worse, indoctrinated with the theories that are treated as "laws", they are doomed to repeat the folly of their teachers and professors. There are many great teachers in this world, but there are also many not-so-great teachers who do not allow students to disagree with them or their beliefs within or without the walls of their classroom. Will such students be capable of making important advances in science? Absolutely; but those advances are certainly going to be limited by their utter adherence to the inaccuracies mathematically built in to the equations used in their field of study that were taught to them by their teachers who are teaching erred theories as if it is fact.

I respect any teacher who teaches a theory as a *theory* and not as fact. Using the term "theory" is, typically, either misunderstood or misused, and often, it's both. For instance, in "music theory" exactly which parts of the study are actually specifically theory and which parts are actually clearly identifiable and provable fact. Pooling everything together into a lump and calling it "music theory" takes away from the idea of a "theory." The same is true of ideas such as evolution and big bang. Evolution is a classic example of this. To the evolution sect, the term "theory" changed from a reinforced hypothesis into a law, but it is still affectionately called a "*theory*." Yet, there are too many unanswered questions to say evolution definitively occurred as stated. This misleading practice presents the danger of creating science zombies; those who can only follow the commands of their masters. "Professor said so, so it must be true." Our minds are a beautiful thing and we should exercise our ability to think freely and seek out truth. Steer clear of absolutist scientific dogma.

If we truly want to be different we must open our minds and allow ourselves to think contrary to conventional wisdom **_when_** things do not add up. We see youth trying to think freely when they are trying to be "different" for the sake of being different. But this is not different at all because they're usually copying someone that they unwisely admire.

At the turn of the twenty-first century, science was headed back to a—Church style action of imposing the pre-Copernican model of the solar system—onto the people and science. We must rebel against these impositions, not frivolously, but thoughtfully and for good and serious reason. The adage "If it ain't broke, then don't fix it" typically is used in regard to absolutist science and it trips us up miserably.

There are a of couple major reasons that the adage occurs: First, the financial and scientific investment in any current theories and equations are enormous, and a great deal of testing relies upon these equations and the assumptions surrounding them. They have become an international-standard including errors and all. The ramifications of having to adjust a close, but incorrect, equation will alter computer programs, text books, and much more all around the world. The second is the social investment; people have built relationships, careers, financial standing, and their status based upon their own scientific models, that are all built upon inaccurate hypothesis, theories, and laws. Whenever a new discovery that defies conventional wisdom is released, such as Copernicus' findings, then the number of people who need to recant their beliefs and theories is enormous. Any changes will cause major international scientific ramifications and shatter the blind scientific faith that far too many share.

Calling the old hypotheses, theories, and laws wrong, incorrect, or inaccurate is somewhat unfair because they do function adequately in many cases. Calling them incomplete is more appropriate; and doing so will alleviate the dangers of absolutist science.

Relativity

Both of Einstein's theories, *Special Relativity* and *General Relativity*, have problems that have been explained away by bending time and space.

In the relativity theories, a moving rod mathematically shrinks when moving along its plane. While this could be so, it does not mean that space shrinks. This goes back to the tangible versus the intangible, and to the definition of "space." If a moving rod shrinks in special relativity, then has the original space that it resided within shrank along with it? Many subscribe to the idea that space is somehow changed in that scenario. This is not likely, but it is how this theory is typically received.

Relativity also implies that a moving clock will slow down. This may also be true, but this does not necessarily slow time. Here again, clear definition of "time" is in order, which we will discuss in detail later. A mechanical clock could behave differently than an atomic clock in this regard, because different aspects of the mechanics of each specific mechanism in the clocks may be affected differently by the linear motion.

In General Relativity, we are led to believe that space-time curves near a gravitational field and that light has a variable speed having to do with the gravitational field. Yet, we say that light is constant.

It sounds as if I am trying to crush relativity theories, but I am not. The inconsistencies are largely from the way the theories are interpreted much the way the Bible has a multitude of interpretations that are all dependent upon *who* you speak to. With the Bible, people tend to interpret it to suit their own fears, needs, wants, and desires, but that does not make their interpretation accurate or valid. This is no different with people's interpretation of the special relativity theory, or the general relativity theory.

Always try to get back to what *is*. There is only one reality and that reality is what *is*. However, our interpretation of that reality is an entirely separate issue, and that is where we trap ourselves.

Whose Reality?

Reality... my reality, your reality, quantum physics' reality, metaphysics' reality, theoretical physics' reality... whose reality shall we use?

Personally, I reject the belief of multiple realities, and I reject that we can create our own reality that is somehow separate from everyone else's reality. I *do* believe we can create our own circumstances, and that what we create we must live with. For instance, if you punch your best friend in the nose, you have changed your circumstances and your best friend may no longer want to be your best friend, and, in turn, your friend may choose to return the favor. Thus, you have changed your reality *and* your best friend's reality together, and everyone else shares that same reality; but it likely does not impact others if they are not in your circle of people. The "reality" that we can change is the only reality that there is, and it is all one reality. It is our circumstances *within* that reality that we actually alter.

I presently reject and am reluctant to follow any belief within the realm of physics that a parallel alternate universe exists where we are simultaneously living a different life than the one we notice here and now, where there are duplicates of each of us making slightly different choices. There is no evidence of that whatsoever, and until someone produces even the slightest of evidence, I will remain skeptical. Alternate universes are a fantasy sort of physics, and are not something I am willing to spend much time thinking about. However, I will not dismiss that possibility entirely, and if evidence is forthcoming, when that accurate evidence comes, then I will be amongst the first to embrace it.

A part of the idea of alternate universes is derived from early twenty-first century quantum physics that had us believing, that from our incomplete tests, that a particle can be at two places at once—for instance, emitting single photons during the double-slit light experiment (Figure 1 Double-Slit Experiment, Page 50.) When observed, the photon only went through one slit, but when not monitored it appears to go through both slits.

This short sighted analysis is highly speculative, because if there is actually a state change in the photon, then when specifically does the light photon change state? Is it before, or after, passing through the slits? And did the photon state change as it was being observed? Did the monitoring device emit some type of interference? It has to somehow interact in order to know if the particle passed by. The experiment, as presented, is far too incomplete to draw accurate conclusions, due to the lack of information pertaining to the above questions, and thus, it is not truly ready for public discussion; nor is it ready for the insistence of "the way it is." There are answers to this problem which will not be discussed in this book due to space constraints.

Anytime we take an incomplete experiment and present an anomaly as accurate, we then walk a slippery slope to absolutist-science. This sort of pseudo-science leads to poorly founded theories about wave-particle duality and allows the belief that the particle was actually in two places at once, or neither place at once, or either place. This opens the door for the multiple universes theories, and for a multitude of other cascading theories that people are actually building their reputations and lives upon. In the end, it allows for hocus-pocus multiple-reality beliefs, allowing us to ask "Whose reality, yours or mine?", and here again we must define *reality*.

Reality implies "*real*", and, as far as evidence has shown, there are only *potential* realties in mathematics. Those potential realities are in the minds of those who originally imagined the equations and are based upon antiquated equations that have long since served their purpose.

We should not be opposed to these theories, but, rather, opposed to the way that they are often presented to the public. It's fun to toy with the idea of an alternate universe, or even an alternate reality, and we should not stop anyone from researching such phenomenon; who knows, maybe they're right. Though, for my part, I would rather see that we understand everything *properly* so that we can actually move to the next level in *true* science.

Here is a reality: While we have accomplished much while using current science "laws," almost no *major* provable advancements have been realized in science since the early twentieth century. It seems that the likes of Copernicus, Galileo, Newton, and maybe even Einstein, are rare. When we allow preposterous theories to go unchallenged, then the cascade of effects from those preposterous theories will affect the reality of everyone around us. This is a reality that we all share!

Chapter 7

Is the Universe(s) Expanding?

Is the Universe expanding? Just how many universes are there? And, is there a multiverse?

Our thinking and beliefs have changed a great deal since the days of the flat-Earth theory and our outdated Earth-centric perspective. Now science believes that the Universe started from an infinitively small point (but some say the ball of matter was the size of a basketball) and that it exploded and has been expanding for billions of years. There are others who believe in a static universe, and there are some who believe that the Universe is actually contracting, so depending upon who you talk to, the jury is still out on an expanding universe.

In the days before what we consider to be "modern science", where we, today, believe they *only* had the six day creation belief at that time, many people just accepted that a God did it all, and they didn't bother splitting hairs on things that they couldn't explain. Everything was attributed to some creator, who we humans believed that we were, or are, too stupid to understand. Things have changed since then. In our open-minded scientific approach we have embraced long-age theory and broke free of the Bible. Or have we?

Here's a myth for you to consider:

Once upon a time, long ago, in a far-away land, Fizeau said, "Let there be red shift," and it was good. Hubble took the red shift and separated the red shift from the not red shift and it was good. Then Hubble said, "Let the red shift show the heavens expanded," there was the hypothesis and theory— Hubble's Law. Then science said, "Let the red shift bring forth moving bodies in the heavens," and it was so. Science put the bodies in the heavens at great distances and it was good. Then science separated the beginning from the end and it was good. Science called the separation the big bang. Then Darwin said, "Let the oceans be teaming with creatures both big and small and let them bring forth every living creature to the land," and it was good, and Darwin called the creatures "evolution."

This seemingly mocking dissertation of creation is as ridiculous as it sounds. This is essentially the sort of thing that has occurred in science during the nineteenth and twentieth centuries. This is not to say that the theories are fables, as many believe the Bible to be, but placing it in a form such as this does shed new light on our willingness to blindly accept big bang and expansion theory as true.

Again, it is important to note at this point, that this is not an argument about their observations, but rather our conclusions about their observations. For instance, I do not doubt for a moment that people actually see something when they report a UFO. The term UFO is actually quite accurate; after all, if it is in the air, then by definition it is *flying*. And if we can see it, then it is likely an *object* of some sort. And since we do not know what it is, so it is definitely *unidentified*. All of the above information is accurate. However, it is our conclusions that lead us astray; we do not know what a UFO is. Most UFO sightings most likely have simple explanations that would make us feel embarrassed when revealed, and, likely, are *not* from outside of our world.

In regard to our perceptions, take, for instance, retrograde planetary movement as observed from earth. (Retrograde planetary movement: the appearance that the planet is moving backwards in its orbit relative to the other celestial bodies.) We cannot deny the observations of retrograde planetary movement. But, the causes for retrograde planetary movement stated in various hypotheses, prior to and even after Copernicus, for that to occur are another story entirely. It is just good science to deny the unrealistic theories they used to explain what they did not understand during that era.

Let us make no mistake about it; our contemporary science is no different now than in any past time. What we *are* witnessing, and what we *think* we are witnessing, may be two entirely different views. *Reality* versus *speculation*—this is very important to remember if we want to progress scientifically.

Past hypotheses have included beliefs such as: If the Universe is infinite and the stars are infinite in number, then we would see a solid blanket of light everywhere we look in space. This is because our line of sight would eventually cross the path of a distant star. Using this same logic, we could then reason that nearly the entire space should be black because our line of sight would eventually encounter space because space is infinite.

The above narrow view of a blanket of light, stated by the German astronomer, Heinrich Olbers, fails to consider two key issues. First, using the same logic, the heavens are mostly empty with regard to the stars filling space, and space has a fairly consistent, infinite amount of non-lighted area. Secondly, we do not know the ability of light to travel infinite distances. We make the wrong assumption that light will travel without end. It may travel infinitely, but there are other more complex factors to consider. Further, our ability to detect light that is too far away, with our eyes or other detection devices, may be impossible due to the low amount of residual energy from the emission when it finally does reach us.

A Fairly Good Estimate of Light

Whether or not we realize it or accept it, science is about index and nothing more. The entire world uses the scientific indices to improve our day to day life. We obviously do not need science to survive or we would not be here today. If all of humanity were gone from planet Earth, it wouldn't take long for the effects of weather to destroy and hide all of our crowning structural achievements. It seems that rock, which is natural to Earth, is among the best building substance for longevity, as is demonstrated by the pyramids of Egypt. It will be several thousand years before we know if some of our reconstituted Earth materials that are made into man-made devices for measuring will stand the test of time equally well.

Science is a privilege. Science is the opportunity to try to know and better understand something. Every time a brave soul courageously stands up and defies the consensus of his or her peers and acknowledges that there are unexplained and ignored errors, they run the risk of being socially destroyed by their arrogantly stubborn peers. That is, of course, unless and until they can properly impress their vision upon their peers. This sometimes occurs long after the forward-thinker has passed. If the courageous soul does have a good hypothesis, then it will often be stolen, modified, and perverted by more prominent and aggressive peers. At some point, the new perverted theory spreads like a disease because it appears to explain observations more reliably than the former erred hypotheses, theories, or laws did. But, a slightly better apparent explanation does not necessarily indicate that it is *accurate*.

Since light travels at the speed of light, and we have a fairly good estimate about that speed, we must *assume* that it is accurate—the speed of light is its own index! This presents a problem when dealing with measuring time and distance using light as a measuring stick to measure a Universe that is hypothetically expanding. Since light is its own index for speed, we cannot detect whether it changes when using light itself to measure its speed.

It is believed that light, its properties, or its mathematical description are somehow a grand unifier that explains all things, but is it? Is light the beat-all substance that is supreme in nature? My guess is, no. I have a great deal of respect for light, because what we call the visible spectrum is what allows us to behold the beauty of the Heavens and the beauty of what we have around us here on Earth. I also respect electricity and gravity and magnetism and life and people and...

This does not mean that what I see is always perceived accurately by me, nor does it mean that what you see is always perceived accurately by you. My personal assessment of the surroundings may not be exactly as they appear. Our assessment of time, space, and the speed of light are, at the very best, educated guesses that are built upon certain assumptions about light and the properties that we believe light exhibits. Our estimate of light traveling the distance of 186,000 miles in one second is based on early calculations, and has been subsequently reinforced by radio transmissions to satellites and other spacecraft. But since light is what we use to measure and is also our index, then if light's speed did change, a planet such as Mars could be somewhat closer or further in actual distance than we believe it to be. Yet, in the end, that doesn't really matter. What matters is that a mission using light as a measuring tool is able to be completed successfully.

If our calculations had a wide margin for error, then it would be difficult to land a Rover on Mars. Being off in our measurement methods would have had the Rover overshoot Mars and be unable to recover from the erred trajectory. Light is a very stable index in comparison to anything else we currently know, which is why we use it; but that does not mean light, or rather our perception of light, is infallible, especially when trying to judge expansion or contraction of the entire Universe.

Light's speed is measured in highly finite experiments; and based on those very finite experiments we have accurately verified, what we refer to as, the speed of light. Again, in a relative setting, light is said to remain constant, and space and time can change—but does this make sense?

Association of Light and Time Travel

Because we can't exactly take our measuring stick out and measure light's distance yard by yard or meter by meter, we need to make assumptions and extrapolate the data of our assumptions to calculate the distances that we cannot measure with an actual measuring-stick, thus, light-years! There is no controversy with the basis-math of a light-year. (Distance light travels in a second) x (Seconds in a minute) x (Minutes in an hour) x (Hours in a day) x (Days in a year) = (A single light-year). Our problems begin when we make certain assumptions about light's peculiar nature.

Time travel has long been associated with the speed of light. Light's speed, or rather its distance traveled in a year, is only a measuring instrument. It is the ruler which we humans use in order to measure vast distances, but it's not quite that simple.

To illustrate, let's imagine light as a very long ruler and compare it to a standard desk ruler. But, this does not work well since light doesn't have the same concrete nature about it as a desk ruler does. Another factor in using the ruler of light is elapsed time. This is something that causes confusion in our minds when analyzing speculative measurements.

Any relatively small scale measurement will be plenty accurate, and any error is reduced to a point of irrelevance when measuring with light on such a small scale. This goes back to the practicality of use. If I am baking a cake, then a kitchen measuring cup is accurate enough. If I am measuring a room for carpeting, then a yardstick is accurate enough. When we measure almost anything here on Earth, light's consistency is so stable in comparison, that any errors are essentially undetectable.

However, this changes when we're discussing greater distances. The greater the distance, then the greater the error, thus, compounding the problem. Since *time* is involved, our perception of time's role when measuring in light-years is vital. Light's apparent properties are the *"Bending The Ruler"* of which I am mostly referring to in this book, though bending the ruler does apply to all scientific indices. If light changes, as witnessed by the observer, based upon the speed of the object emitting the light, then judging the Universe to be expanding is highly speculative. We can say red-shift is evidence of an expanding Universe, but we truly do not know how fast the Universe is supposed to be expanding.

Doppler and Red Shift Study

The Doppler Shift (briefly explained later) is proposed to behave the same for all waves including light. However, this does not mean that this is completely accurate. Observations appear to indicate that the emitted waves are compressed when a wave emitted from a source is moving towards an outside observer. This can be demonstrated in a pitch change when a vehicle is approaching you compared to when it passes you. It appears that this frequency change is likely so with radio and light waves as well.

There is a "however" in this; however, there is an amount of uncertainty pertaining to the speeds of various "wave" frequencies that are detected at tremendous distances. Making the assumption that light red-shifts, from the recessive motion of a luminous body from the observer's perspective, is speculative. This speculation causes other problems and discrepancies which tend to get ignored.

Our insistence of the constancy of the speed of light, or maybe better stated, the rules we apply to the speed of light, should put up many red-shifted flags. Since our repeatable experiments can only speculate what is actually going on during the experiments, it means that what we believe is going on is our best guess. Without getting into petty debates on theory, if you putt a golf ball into a cup, then you have seen it actually travel to the spot and drop into the cup—you have witnessed it! However, dealing with phenomenon experienced with the electromagnetic spectrum is somewhat different. This is because once we get beyond the furthest point ever traveled by man-made craft, we have no proof of accuracy: the further the distance, then the greater the potential error.

Many of the explanations formed over time are likely to be proven somewhat accurate and are extremely useful to humanity for practical application. Yet, we are still left with the reality that most of our methods are only best guesses, thus, leaving it all open for rational debate. The problem is not the "best-guessing" part. The problem is our refusal to look at conflicting information. By conflicting information, I am referring to information that does not fit with our current understanding of the light-speed model to a point of unwillingness by the scientific community to consider *other* possibilities.

The science world should not stop in its tracks and suddenly change direction when new information is presented. However, there is far too much utter refusal by the science world to consider anything outside of conventional scientific wisdom. This "conventional" wisdom often is not completely accurate.

If consensus is our ruler, then based upon "conventional" wisdom, everyone should accept that there is a Creator because there is a globe full of people who say that it is so. When in question, the believer should wonder if there is a Creator, and then find the evidence to support their claim, and science should do the same. Red shift may truly be from celestial recession, but it may also be that it is not from that. This is something that should be thoroughly investigated.

There are serious discrepancies, which we spoke of earlier, with regard to the age of the Universe and with regard to the relative positions of the celestial bodies within the supposedly expanding Universe. Unless we want to accelerate all known matter at enormous speeds of unrealistic multitudes times the speed of light, we cannot say that the Universe is caused from a big bang. It appears that the big bang requires an exponential amount more of blind-faith than believing in a Creator does.

We can imagine that what we call our "laws" of physics did not exist before the big bang, but that is unlikely. The erroneous black holes in the big bang theory are too numerous to discuss in a single book. Some sort of static Universe theory is far more likely and does not require a Santa Claus-like belief of a big bang that just spontaneously burst forth from a single point of nothingness. Static Universe is far more likely, and it would allow for a galaxy to be seen 13.2 billion light-years away *without* cheating the math or breaking the beloved laws of physics, whereas a single big bang does not allow so without major contradiction and imaginative blind faith speculation.

The appearance of red-shift that is seen is more likely a phenomenon of the magnitude of distance, than it is from the action of a receding galaxy. As discussed earlier, Heinrich Olbers postulated that the Universe cannot be infinite or the entire heavens would be a luminous blanket. This is because every line-of-sight point would cross the path of a star at some point. This is simply wrong for the two reasons mentioned earlier. We can add to those two reasons a third point: It is possible that there could be an infinite amount of galaxies lined up behind one another making them hidden to our line of sight, leaving only the nearest one readily visible.

When looking at the night sky with the naked eye, it seems fair to say that some stars appear far dimmer than others. Hypothetically, assuming the light energy to already be upon us, a galaxy infinitely far away would have infinitely less light energy hitting our eye. An additional point with Olbers' perspective is that there very well may be a dim blanket of light that encompasses all that we see, but we may not be able to detect or realize it, both, because it is very low energy and because we are immersed in it and have become accustomed to it as "normal." This is similar to being in a pool of cool water and becoming accustomed to it. Everything that we detect automatically inherits the error factor of ambient space-light-energy unknowingly built into the equation because it is an inescapable aspect of space. We do not know and cannot prove zero energy because we have no true index of zero energy.

The Universe could possibly be static and it can be infinite and it can be either newer or older than 13.7 billion years. It won't damage practical use of our current laws of physics, though it may damage a few egos if it is not as we currently believe it to be. If light red-shifts, then do we know for certain exactly how much it shifts based upon any given speed of departure?

Is $e=mc^2$ Inaccurate?

If Einstein's $e=mc^2$ formula is inaccurate, then our estimate of what would happen when the mass of an asteroid strikes another celestial body will be incorrect or inaccurate as well. Our inability to see or measure the error will increase with the magnitude of the action. How do we measure such an action that will put an end to us if we are too close to it when we observe it?

When multi-megaton bombs are detonated for testing, do we know for sure, *exactly*, how much energy is released? No, obviously not, that is pure speculation. Our scale is based on tons of T-N-T. I can only speculate here because I could not find documentation, but I feel quite certain that it has never happened that multiple _mega_ tons of actual T-N-T have been detonated simultaneously at the same location in effort to calibrate our *estimate* of atomic power. Though it is known that in the 1940's 100 tons (0.1 kiloton) of T-N-T was detonated as a means of estimating the power of larger nuclear devices.

Our assessment of the power of a nuclear weapon is based upon the $e=mc^2$ principle. While there appears to be little question that there is an association between mass and energy. Placing the c^2 in the equation is the equivalent of saying e=m$^{\text{really really big number}}$. There is no question of the broad destructive forces of a nuclear weapon, but the accuracy of its effect will decrease with the estimated magnitude of the weapon. What we do know is that such large explosions are very good at breaking a lot of stuff very quickly.

Since a nuclear weapon is dealing with only a tiny amount of reactive material, it is no comparison to the calculation of the true energies involved in a star, such as our Sun. And our Sun is no comparison for the calculation of energies in a galaxy, such as our own galaxy. And again, our galaxy is no comparison for the calculation for energies required to form the visible Universe, and the *visible* Universe is no comparison for the calculation for energies needed to form the *infinite* Universe. The magnitude of error is off the charts and will remain so until we actually get the information correct. But, with our current habit of bending the ruler, it is highly unlikely that we will ever be able to reveal what is true, because our minds are currently fixed on inaccurate information.

Depending upon how we use a ruler, a carpenter can build an entire home with a ruler that is off one percent with little or no ill effects. Since our size standards for building homes are well set, we can change the carpenter's ruler one percent and he will mostly not notice it. He makes adjustments for the anomalies that he sees by stretching space at each important point, such as where universal items like doors and windows are positioned. Overall, the house is going to be good, strong, and successful. In the end, he may have unnecessarily shrunk space just the same as our theoretical equations do. If the carpenter understood that his ruler was off then he would have made appropriate adjustments and the building would have been perfect. If he doesn't understand the problem, then either the building will be one percent off or he'll do things like using more shims in door and window openings to compensate for the discrepancies.

If a ruler stretches or shrinks, it will cause the environment being built that is using it to measure to equally stretch or shrink. When it comes to the *space* that the building is built within, it's different because we are not building space; though, we act as if we are. What we are actually doing is referencing space. If our ruler is wrong, but we believe that it is correct, then we inevitably will have bent, shrunk, or expanded space in our minds and in our minds alone.

Expanding at the Speed of Light

Can something, or can something not, exceed the speed of light? If something can, as is proposed in the big bang, then, according to e=mc^2, big bang science has much explaining to do. If light is susceptible to Doppler Effect, then we cannot be sure of the position of the celestial bodies. And based upon our current beliefs about light's speed and the distances we measure with it, we would perceive a celestial body as much further away than it actually is.

Our use of math has some serious anomalies. If light is constant to the observer, then red-shift may not be able to occur. If red-shift cannot occur, then we must rethink our views of the age of the Universe. Even if it can occur, we must try to understand the anomalies still observed with that explanation. Depending upon who is interpreting it, when we compare Doppler Effect to the pitch change in sound waves, we're making a statement that potentially does not properly agree with the "c" *constant* and its connection to relativity. Our maths are incomplete patterns of unknowns. Using various kinds of math for speculation is fine to do, but accepting them as factual and final is a grievous error.

As per Einstein's theory, if a moving object's time is different when it is in motion, then who is to say whether the object ages faster or slower when in motion? We simply do not know the answer to this; though, we do seem to have a correlation with time and gravity, as is apparent in the clock differences of orbiting satellites.

Who gets to choose what is moving? Does direction towards or away from the observer matter? Whose clock is actually changing?

The motion or velocity portion of gravity and mass is greatly misinterpreted. Depending upon the speed at which another celestial body is moving away from us, the perceived duration of the movement from a start-point near to us, to the point we currently observe it at, can be greatly altered by the opinions regarding the "c" constant and how it actually behaves.

According to relativist mathematic theory, a body moving at the speed of light won't age. Therefore, something that is 13.2 billion years old and 13.2 billion light-years away has had a velocity of the speed of light for 13.2 billion years at minimum. This means that the object is not old at all. In fact, according to the theories proposed (built upon the shoulders of $e=mc^2$) the distant galaxy has no age at all. This, of course, all depends upon which body is moving—in theory.

By now you should be beginning to be able to see the inconsistencies that we have built into our mathematical equations that deal with the origins of everything. From an alternate perspective, if the Doppler Effect works with light the same as it does with sound matter, then, in theory, a receding body moving near to the speed of light would appear to be considerably older or younger. This is depending upon our interpretation of the information coming to us from that moving body. Just as a sound wave lowers in pitch, so, too, is light believed to change as per the Doppler Effect. So, what is the effect of that regarding aging or time?

In a light-beam timeline, all emitted light information that we receive from a receding galaxy would appear to be moving slowly and take very long periods of time in order to accomplish a given end.

A "light-beam timeline" could be thought of as a movie film being emitted towards your eyes (Figure 8 Movie Projector Timeline, Page 176.) The movie's apparent play-speed will be reduced proportionately as a percentage of the speed of light at which the source is receding from your eyes.

Since the distance across the distant galaxy is believed to be extremely vast, the relative movement within that distant galaxy will be undetectable and therefore unobservable to any practical point. Additionally, if the galaxy is moving away from us, then our perception of the movements would be that they are slowed considerably; thus, to us it appears static in reference to itself. Since we have no hard tangible data with regard to the actual distance of a distant galaxy, all we can do is to make wild and random speculation based upon our current understanding of light's speed.

Because we are very self-centric beings, we reference everything that we experience from **our** current understanding of **our** own experiences. If a body is receding from us at near to the speed of light, its internal movements and its recession may appear to us to be much slower or even motionless. This means that when we look at it we will believe it to be at a very different location than it actually is, or where we calculate it to be.

When we imagine our galaxy, we picture it to be a glowing disc of stars heavily concentrated. This, though, is simply an incorrect perception. Stars, relative to their distance from us, are far smaller than they appear to us. When we view something very small, our perception comes down to glare, and pixilation in our own eye as well as in the cameras taking the pictures. Much of the glow may, in some cases, also be attributed to "space dust" or gases or other radiation. Our fantasy-based, computer-animated travels through space are done at speeds that are so many times faster than the speed of light that it is difficult to comprehend. Passing multiple stars in any one second would put the speed of the fantasy trip at something near 250,000,000 times the speed of light. Passing any galaxy in a few seconds makes the speed many times more than that. I enjoy fantasy rides through space via computer animation, but our real world science psyche that is often based upon such fantasy is irrational, which is where the ruler bending problem often occurs.

The glow or glare that we see around a star gives it the appearance that it is larger than it is in reality. What we believe to be other galaxies, are full of stars that emit light. Each star's glow, illuminated "space dust", gases, and other radiation makes it look like these most distant galaxies are glowing and filled almost solid with stars. But if we were in one of those galaxies it would not have that same feel as it appears in our pictures of them. It would be vast, open, and empty, just like it is with our Sun and the stars near us in our own galaxy. In reality, the galaxies do not glow in that way (with the exception of various gases and space debris, and also radiation that is often not visible without special equipment that can detect frequencies of light or radiation outside of the human visible spectrum). The distances are so vast between their stars that traveling from one star to another star at the speed of light would likely take many years, even tens of thousands of years. We believe that it would take tens of thousands of years just to cross our own galaxy at the speed of light. The day to day progressive movement of anything moving at less than the speed of light will be almost undetectable to us, and any minute changes we might perceive are likely to be unreliable at best. This is especially so because all stars in a particular galaxy are likely all flowing in the same well-synchronized direction.

The sci-fi entertainment perceptions that we now have are disconnected with what we are actually capable of with our equipment. Similarly, our actual scientific perceptions are also disconnected from the scientific math and sound logic. We should try to keep it all connected so that we can better relate it all, *and* so that we can better separate fact from fiction.

If any of the aspects of Einstein's theories are incorrect, then we must rethink our estimates of the age and size of the visible

Universe and develop a new perspective. We cannot continually disconnect various aspects of the theories, and then apply only the parts that *we* want, in effort to make our calculations work. Our perceptions are very convoluted and we are demanding to have the lights both on and off at the same time. To this, a theoretical quantum physicist might say, "but we **can** have the light both on and off at the same time."

Consider the speed differential perceptions that Einstein spoke of where our perceptions of the motion of the moving person as received as a result of the emission of light from the moving person's source of movement and light emission. Since we cannot see this sort of motion in a distant celestial body to any usefully measurable degree, we have zero reference to know for certain that this is the case. The reference that we do have is what we believe to be a year and the theoretical distance light travels in a year (a light-year.) This reference is in relation to where we think we see a distant galaxy.

If a big bang did occur and the expansion was anywhere near the speed of light, our perception of the distance may well be far greater than it is in reality. Based upon this we could age the Universe anywhere from several thousand years to billions or even trillions of years. This duration of time will vary depending upon exactly how we choose to bend the ruler. Our assumption of the constancy of light is at the heart of this anomaly.

Along with this light-speed issue and Einstein's imaginary rubber-sheet-plane style gravity-theory (Figure 7 Rubber-Sheet-Plane, Page 119) we also have to consider our understanding of the half-lives of various atoms. (Einstein's imaginary rubber-sheet-plane style gravity-theory is where gravity is compared to the depression that would be made if a ball were pressed into a sheet of rubber that has been stretched flat and tight.) If an atomic clock is affected by its position in orbit, then we cannot assume any dating to be reliable when considering long-age calculations based upon any sort of atomic scaling because doing so can cause us a great deal of mental deception.

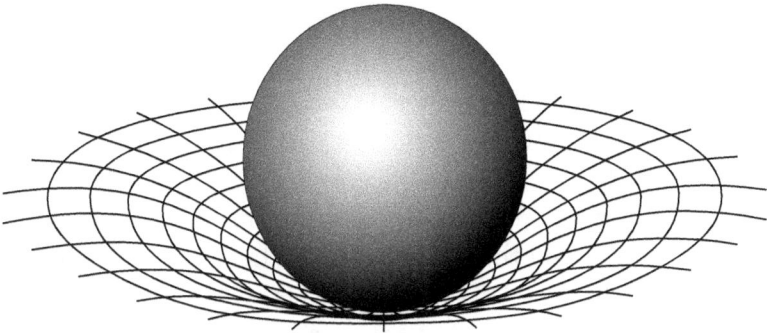

Figure 7 Rubber-Sheet-Plane

We must force people to answer for their own contradictory theories that they are proposing and perpetrating on society, especially while accepting public funds, and even more so when glaring anomalies exist within those theories. Science has embraced too many theories as *truth*, even though there are obvious problems within some of these theories. These problems can become even worse depending upon who is interpreting the data.

Time Cannot Change

This section will cause confusion depending upon what your definition of *"time"* is. *Time* measurement is an arbitrary idea because it has no basis, but we believe it to be something that is tangible.

We use a ruler to measure distance, we use a scale to measure mass, we use a thermometer to measure temperature, and we use time to measure time. What? Let's back up a bit here, just what is time and just what do we measure it with? We believe that we measure time with a clock or a timer, but are we actually measuring time? No, if that were the case then we could turn back time by simply turning the clock backwards by moving its hands in the opposite direction.

We do not measure time, we count in segments of time. This may take the romance out of the idea of the dimension of time for some people, and it is *time* to accept a simple truth about time. Time is not substantive. Time is like the idea of distance. When we measure distance we do it in increments called inches or millimeters. We break the designated distance into segments of a given length, and then we count how many segments there are in the specified distance. We can use this method in reverse and plan ahead, and then count out how far we want a distance to be.

Time is no different than distance in this way, but because time is a bit more arbitrary, we tend to miss a key point: we don't actually measure time. Just like distance measures space in increments of a chosen size, time measures something in increments of a chosen time length or duration.

With distance we can measure the distance from one wall to the next using inches and millimeters, or we can measure from one city to the next using miles or kilometers, or we can measure from one star to the next using light-years. And with liquids we measure in gallons or liters. But, in each case we are measuring *something*, usually either the mass, or the space that a substance occupies.

Volume is to *water,* as time is to... what?

When we measure something we do not break the actual item up into units, we break an aspect of it into units of measure. The "measure" being: length, width, height, weight, volume, temperature, etc...

The three primary size dimensions, being *length, width,* and *height,* serve our purpose well because they can tell us how much space something takes up... or can they? The concept of three dimensions is very short-sighted. If I give you three dimensions, can you tell me the volume of the vessel? No, absolutely not. Length, width, and height tell us nothing of the shape of an object.

When we believe time to be some mysterious fourth-dimension, then we have blinded ourselves from reality. When we try to imagine that more dimensions of some mysterious manner exist, then we are living in a fantasy world. We claim that there is no "Heaven," but then claim that there are parallel universes where we, or our twin, live different lives. This sort of conflicting thinking is not very scientific and it does little good to advance science.

Again, let us consider *length, width*, and *height*. If I give you the primary dimensions of a pyramid, can you tell me the volume? No you cannot, not without a great amount of uncertainty. "Why?" you wonder. Ask yourself this, how many sides does my pyramid have? Without this information you don't know what equations to use and you do not know if the planes of the pyramid are flat or if they have any curvature, or even if each side has the same dimensions as the others.

There are certainly more than the three dimensions of length, width, and height. Weight or mass is the first that comes to mind. Now that we have an object that has size and mass, we can consider shape, and after that, we can consider time, right? No, not exactly.

Volume is to *water,* as *time* is to???

Start it this way:
Gallon is to *volume,* as *hour* is to *time*
Volume is to *water,* as *time* is to... what?

Time is a measurement of duration of something. When we measure time, we are measuring how long something **was**. We measure from one moment to the next. We measure from this moment... *pausing...* to this moment. Duration is somewhat of an abstract idea. If you recall, according to Webster, something that *exists* is "*tangible*", thus it can be detected, but "*time,*" as we call it, cannot be detected, yet somehow we can measure it. We also cannot detect distance or weight, but we can measure them.

Volume is to **water** as **time** is to... **existence** (or how long something has been or will be).

For all of known history, our original clock that we used for marking time was a single rotation of the Earth; longer periods of time are measured by the Earth's position relative to the Sun and other stars—the ultimate clock! But again, recall that we are not measuring time any more than we measure distance. Time (or duration) and distance are all ideas. Duration and distance are nothing; instead, it is *what* we measure when using time and distance that is important in this regard.

Time, as we call it, (which we will explore in-depth a bit later) cannot change, but the counting of it can. When we adhere too strictly to potentially inaccurate but usable formulas or equations, and we cannot yet imagine or understand in our minds how to deal with an anomaly, then we cheat by bending the ruler. Bending the ruler in theoretical mathematics works to theoretically explain the observations, but that does not make those theories correct or accurate.

At some point, reality steps in and what *is*—is what *is*. Just because we currently cannot explain what *is*, does not mean that we must assume that something exists or occurs which runs contrary to common sense or reality. Because of our potentially skewed understanding of the cosmos due to our acceptance of potentially flawed theory, we have decided that space is bent and is expanding, when the more likely reality is that space is static and it is not and cannot be bent.

Chapter 8

Matter

What is matter? What is this stuff that all things seem to be made from? What is our definition of matter? **Matter** is defined as a *physical substance*, but who gets to determine what a "physical" substance actually is?

If we're not able to see something, then can we claim it to be a physical substance? Using electron microscopes we can see atoms, so this offers us reasonable proof that atoms exist. But the matter of *matter* at the quantum level is considerably less tangible. Once we penetrate the subatomic-particle level, then all bets are off; but while we have not yet been able to fully determine the specifics of the subatomic level, we should still keep trying.

The deeper we look, then the more the tangibility of what we are looking at vanishes. But, does that mean that something is not there? Tangible or touchable is an elusive area of study because there is no point of reference established; and our ability to establish that reference point can easily be moved by bending the ruler.

Our evolutionary-minded experience in life has been both speeding up our understanding of our environment, and hindering it at the same time. The open-mindedness of *not* deciding that **our interpretation** of Genesis chapter one is "the way that it is" has allowed us the freedom to peer into some great unknown areas. However, our insistence that the likes of Darwin and Hubble are nearly one hundred percent correct on all accounts is holding science back a great deal, which is something that we should take considerable issue with. It is not an issue of whether or not there is a Creator, but rather a narrow-minded insistence that we currently are understanding the whole picture because *we* say we do. Our understanding is a gift, whether it is a gift of nature or the gift of a Creator, it is nonetheless a gift which has been given to us.

Our quest to "look" for something tangible has us believing that everything is always going to be able to be quantified in the same manner. In some ways this is a ridiculous thought, and it is no different than assuming that we can look through a wall with common corrective lenses. When we have a scientific hunch, we must follow those speculations to see if they prove true, even if they are only theoretical and nearly no evidence has been found about them. The problem occurs when we are too rigid and refuse to alter our theories to accommodate our actual findings.

We can look through a wall, for instance, with the proper tools. This is similar to the concept of teaching peek-a-boo to a child. To an infant, when something is not seen, then for them it is irrelevant. It is the parents teaching them that allows the child to understand that the thing is still there. Eventually the child figures out that the thing still exists, and then they attempt to grasp for it. But, for whatever the reason, science keeps looking for a tangible item that may not exist in the way we understand "exist", and because science cannot detect it, we then say that it doesn't exist.

What is Matter Created From?

At some point and level, we will eventually come to the understanding that everything is made from nothing, or in a more scientific term, it is matter-less. Our finite minds have built false barriers for us that we have been unable to break through over the thousands of years of recorded human history. Because we insist upon always looking for the origins of the Universe in the same way, we have barred ourselves from seeing the truth. Slowly, over the millennia, people have broken through various barriers which allowed more articulate interpretation of our collective observations.

The ancients could have done what we do at any contemporary time if they had the understanding that we have today. All of the materials were available and at their disposal in their day, but they could not see for the same reason that we cannot see some things—*our preconceptions* of the way it is and our lack of vision for what could be. We lack creativity when we insist on applying the same "laws" or rules, that we have made up about physics, to everything else in the Universe.

In our arrogant human nature we expect that somehow *we* make the rules and that physics obeys *our* laws. This might sound silly, but then, why do we say something breaks or defies the laws of physics. Matter does not defy the laws of physics, the "laws" of physics defy the "what is" part of matter. That is to say that we describe it wrong with our mathematical laws. Matter will do its work whether we are dead or alive. Our man-made rules (our interpretations) of what matter is supposed to do are useful, but they are not laws. They are nothing more than descriptions of our often flawed and lacking observations of matter and all that is made of it.

Early twenty-first century science simply had no clue what matter is, or what it is made from. It is in our nature to want to know where we are from. But if we can't get over our arrogances, then we will remain blind to the truth about what *is*, and we will continue to insist that our laws are true. Our search for scientific answers should never stop, but our blindnesses should stop.

Science stops at the intangible and passes the gauntlet to metaphysics because philosophers that originally created the physical sciences inadvertently divided an inseparable whole. Since this separation, and before, much of metaphysics has taken a Santa Claus-like approach of a fantasyland that does not exist. This does not mean that there is not a spirit realm or a Heaven, but rather that it is often misunderstood. If we say that science proves that there is no Heaven, then the religious and metaphysics group becomes angry; and if we demand that Heaven exists, then the science group becomes angry. We can call metaphysics the metaphysical-sciences to sound more sophisticated, but that will not bring us any closer to actually understanding what is true.

It's difficult to describe something to people that they do not understand or do not know exists; this is due to the fact that they have never experienced or conceived it in their minds. Trying to explain something that is unknown to someone is a truly monumental task. We learn and teach in analogies, and our analogies are relative to each of our own experiences in life. Imagine trying to describe astrophysics to a 18 month old toddler; do we imagine that they will grasp the concepts involved? The same holds true with adults: If an adult has never experienced a brick, then my best hope to explain a brick is to try to get them to combine the ideas of a **square** and a **rock** in their mind. This might help them understand, in their mind, the concept of a *brick*. However, if they do not understand *square* or a *rectangle* and a *rock*, for example, then I'll have a truly difficult time getting them to grasp the idea of what a brick is. The subject of *matter* is no different: anti-matter is a term that is very short-sighted indeed, but it currently may be the best description that we have to convey *nothingness*. Yet, we understand that something more exists. *Anti* means opposed to, but *Ante*-matter is a more appropriate term to use than *anti*-matter. From a definition standpoint *Ante* can mean both before and against, thus giving us a more accurate perception of something yet unknown and not understood.

Ante-matter can come before matter; meaning that it is sequentially first, and thus, it is what allows "matter" to *be* (or to *exist*) to begin with. Other views of matter are opposition theories where there is almost a cosmic fight of *matter* and *anti-matter*. Some of the conclusions of past hypotheses are somewhat metaphysical in nature, but are then cleverly turned into anti-matter.

The term **nothing** is an interesting term because, for most of us, in our minds, "*nothing*" indicates no matter. We typically don't believe that anything can be made from nothing, which causes us to bend our rulers in order to make things fit our imagination, or lack thereof.

Assumptions of Singularity

The 13.7 billion year age of the Universe assumes that all of existence started from a single infinitely small point called "Singularity"; this is an assumption that is very misleading. Our mathematical "laws" have caused us to convince ourselves that singularity can and does exist, and that black holes may have some sort of a role in that. However, we do not have observations of such to prove these hypotheses. This does not mean that we should not propose singularity, but we should open our minds to the possibility that singularity does *not* exist, and that singularity may be impossible no matter what *we* choose to believe and "prove" with our limited mathematics. Singularity is something we use to rationalize what we don't grasp. In other words, the smaller we can mathematically make the Universe, then the easier it is for us to conceive the idea that everything came from nothing.

Many people choose to believe what they want to believe, and will bend the ruler at will in order to force their observations to fit their scientific blind-faith. You must not understand this to mean that science is wrong, but rather our interpretation of what we think we see is often wrong or inaccurate; "wrong" being completely without merit, and "inaccurate" being partially with merit and partially without merit. Inaccurate can be thought of more along the lines of doing the process of something right but making a mistake in the process; where wrong is not even having the procedure correct with which we attempt our calculations. For instance (2 plus 2=something), and (2+2=) is accurate when used for the function of addition; but (2x2=) is not accurate for the function of addition. So while the answer for both is 4, if you assume addition in both cases, then only one of them is actually accurate and the other is altogether wrong and only has the desired answer by mere chance of the particular numbers being evaluated. (2+2=5) is *incorrect* when doing *addition*, but (2x2=4 That is to say 2 *times* 2=4) is *wrong* when doing **addition**.

Singularity theories can vary, but are always mathematical and highly speculative in nature. As discussed earlier, our illusion of the three primary dimensions of spatial relations are: *length*, *width*, and *height*, but they lack the information needed in order to accurately calculate *volume, mass,* or *duration*. Shape, in itself, is a physical dimension. In fact, *volume, mass,* and *duration* are all physical dimensions as far as each aspect itself is concerned.

Mathematics allows us to create dimensions that cannot exist in the real world. In computer programming, we create multidimensional arrays of data. Multidimensional arrays are something that do not fit into the cells of a standard three-dimensional cubic grid structure. However, while we can produce these mathematical structures with the mathematics involved in computer programming, it is still only an *idea*.

The following point is very important to understand: While we imagine these multidimensional arrays to exist in computer programming, they do not actually exist in reality. They are irrational ideas—not reality. Multidimensional arrays exist on a flat computer disc or within the limits of a computer memory chipset, and they are, in fact, not multidimensional in any form. It's okay to imagine that we can have these extra dimensions, but then we truly need to redefine the word *dimension*. In our minds we want to bind the dimensions into a 3-D space, but that is very improper thinking. 3-D space is only an aspect of something (or three aspects); everything has other aspects within that 3-D space, such as color, mass, temperature, shape, texture, sound, and movement, etc.

Imagining that these other additional aspects of data, which can be held in a computer's multidimensional data array, is fine as an idea, but trying to somehow picture each of them as a new point in three-dimensional space is foolish. Structurally, in a computer, a multidimensional array is more accurately pictured as a relational database than it is a multidimensional array. For the purpose that it serves in a computer, a mental illusion of a multidimensional array is fine because it is a good way to relate a single piece of information to the many aspects needed in order to describe it. Dimensions, as we call them, are pieces of information about a physical object.

Singularity can be thought of in a similar manner. Singularity is a mental effort to force all aspects into a single cell or point. *Mathematically*, this is possible to infinity, but that does not make it possible in reality; yet the experiment is a good *mental* exercise. When we get hung up in infinite mathematical equations, then it is as if we are improperly viewing a graph and taking a small piece of that graph and projecting it in both directions with the same inclination angle as the sampled portion contains as is illustrated in the earlier figure shown (Figure 3 Stock Chart, Page 70.)

Taking a small portion of a graph and projecting it to infinity is similar to assuming a child will grow at the same rate for the rest of his or her life. If a child did grow at the same rate that they grew from birth to twelve years, then by the time the child aged to eighty-four years old they would be about thirty-five feet tall. From our day-to-day observations we see this to not be the case and understand that this is an absurd thought. The same is not true for proposing singularity from a single math equation; this is because we simply are uncertain of how things would look due to our lack of observational experience. This is unlike watching people grow where we have witnessed it daily from birth to death. We can have such imaginative speculations in physics because we simply do not know, and we currently lack a sound way of proving or disproving such theories. For good or for bad this allows for some very absurd proposals to be offered.

The Assumption of Constant

Using mathematics assumes that *all* things function constant according to what we have proven with the mathematics. Using current math to calculate gravity too far beyond its scope is possibly inaccurate, and it will allow us to believe in "Singularity" and "black holes." Is it possible to have a "black hole?" This answer will depend upon if light has mass, or more simply put, if light is able to be affected by gravity.

A black "hole" is truly an improper name and only our mathematics allows us to hypothesize that a "black hole" can shrink in size to an infinitely small point referred to as singularity. When we allow this rationale, then we are allowing ourselves to mathematically stretch this theory so far beyond reality that we are able to hypothesize that a black hole can be made from a low gravitational item, such as a rock. That is, of course, only if we can compress it to a small enough size.

This lays premise on premise, and it is a very dangerous way, or maybe better stated, a very foolish way, to use our math. It's a mistake far worse than measuring a mile going end to end with a six-inch ruler. You will have an additive or subtractive error from going end to end with each successive leap-frogging of the rulers, thus causing the error to potentially grow. Math, built premise upon premise, can be even worse because a tangible measuring ruler is constant, where *math*, on the other hand, can become exponential. When an exponential number is combined with another exponential number it allows for some pretty outlandish blind-faith scientific beliefs.

On occasion, we must separate ourselves from our work in order to reevaluate our observations in relation to reality. To assume that light is a "constant" in the way that we do, is both foolish and dangerous, not to mention contradictory. There is much data available that has been collected while assuming that light is constant, and through that assumption "proving" it in our own minds to be so. However, light's constancy is an entirely different issue than our observation of light's speed. For all practical purposes, light's speed is the most constant phenomenon we are aware of that we can effectively utilize for measurement purposes. Its apparent stable nature allows us to measure our local environment with very high precision, and, in the end, that is all that matters to us—practical application.

Speculative distances assuming billions of years are subject to errors of not just a percentage point or two, but potentially thousands of percentage points in error.

It is not possible for light to be constant in any way. Either our perceptions of Einstein's theories are incorrect, or the theories themselves are incorrect. The theories and our perception of them are best used within their scope *in our imagination* and speculations.

Location is Primary of the Tangible

Location is primary to measuring all tangible elements. Without location, matter cannot scientifically exist. All measurements of size will always be from *here* to *there*. Location of volume is not *length*, *width*, and *height*. Location is at minimum three points, or point locations, for a flat plain, and then, at minimum, a fourth for height. Our view of the three dimensions being *length*, *width*, and *height* is grossly short-sighted. When we figure *length*, we need at least two points.

When we figure *width*, we need two more points or we can share one point and then have a triangular plane of zero height.

To give a triangle volume, we need to add at least one more point that is not in the same plane as the other three points are located. Each of the four points is a location in the space of the expanse. Those points can be moving within the expanse, but are stationary relative to one another.

The purpose of discussing these points in space is this: If something has a single point (zero size), in other words, all four points are in the exact same location, then the form has zero size. My question for thought is, then can it exist? Webster's etymology would indicate not. This takes the thought process back to *terminology, definition, understanding*, and coming to an agreement on what we mean when we say "*exist.*"

We could say that we cannot see, feel, hear, touch, or taste an atom, but this is not true. We may not perceive a single atom, but we do perceive atoms in groups. A zero point is different. An infinite amount of whatever is occupying a zero point is still zero in size and would have no detectable properties. So then, can it be considered physical? As discussed earlier, the word **dimension** means to "*measure*" and the **di** part of the word means "*apart*" or "*two.*" It seems fair to conclude that our understanding of this is to *measure* from one point to another. Something with zero size cannot have **dimension** as we think of the actual word, because the points share the same location, or better stated, they **are** the same location, thus they cannot be "*measured.*" Can something exist at zero point size and zero point? To be more specific, does zero point size exist in a physical sense? This question nears to the philosophical thought of: What is nothing?

In our imaginative way, we typically picture the individual components to be little balls of stuff. Once we get below the electron microscope level, our ability to see form is terribly obscured, and determining size becomes considerably more complicated. Finding a location of points with dimensions is even more difficult with the illusive subatomic particles. Because subatomic sizes are so small, we simply do not have the ability to do much more than take educated guesses, which appears to be working so far... somewhat.

Our indexes of time and light have different implications for attempting to assess astronomical distances, versus attempting to understand the inner workings of an atom. However, the theories that are questionable will affect each area of study, and the level of impact will be determined by how far we are willing to bend the ruler in either instance.

A major problem with outer space is distance errors. But with particle physics the problems arise with our expectations of the theoretic durations and sizes. Our expectations are greatly affected by our bent rulers. These potential errors are ultimately combined when calculating the mass and size of black holes (better described as dark stars, or hidden stars), or when calculating any other astronomical phenomenon.

Disorder from Order

Our understanding of all of Creation is greatly affected by our understanding of reality. The reality I am speaking of is not the reality of Creation, but rather the reality of our measurements used in our sciences.

Our contorted and bent ruler has been leading us astray, and we are being lured by money and scientific status to our own demise. Personally, I believe that while Einstein would have fought for his theories, he would also have conceded when a *better* explanation came along. And I also believe that, until something better came along, he would have admitted to the potential errors shown by certain anomalies in his work.

Einstein believed in a god, and in his mind he knew that his "God" would not create arbitrarily. This led Einstein to his conclusions from which he derived his theories. In essence, what

Einstein was saying is that there is order in the Universe and his God made it that way. Where there is order, there is pattern. And where there is pattern, things *can* be predicted.

The quest of science has been to try to explain the patterns that we see and experience. However, our bent ruler has been a hindrance to us because we are essentially saying that we see a pattern, but then we demand that no pattern exists. We do this by bending the ruler, such as saying light is constant, but then utilize the Doppler Effect and red-shift to see what we believe to be a detectable speed change.

In reference to patterns, small orbs will make a different pattern than large cubes. Trying to apply the same mathematical explanation for a small orb to a large cube is incredibly ignorant. Here again we must understand the word **ignorant** to be one who **ignores.** To **ignore** means "*in*" or "*not*" plus "*know*." When we ignore something, we are often choosing to **not** know something. This is usually due to our internal frustration of not being able to comprehend what we are actually looking at. Ignoring anomalies and explaining them away by bending the ruler is very unscientific, but is common because it releases our minds from that which we cannot yet comprehend.

As mentioned earlier, matter is something within space and time, meaning that within the expanse it occupies both location and volume, plus it has *duration*. With zero duration it does not exist because *it is not now* and *was not ever*. Our dividing line between matter and that which is not matter is very blurry; and this is where most of our scientific ruler bending occurs. This Universe has repeatedly proven itself to us that it is reliably ordered and chaos *does not* exist within it. "*Chaos*" is a word that is both misused and misunderstood, and so is the thought that is typically attached to it.

We believe *chaos* to be "*disorder.*" The word *order* is, for some reason, intuitive, and we grasp the fact that *order* is, in essence, an *arrangement*. Calling *order* an arrangement has theological implications, and since this is a book about science, we will leave that aspect be for the most part. If something is not *ordered*, then it is *disordered*. Order is the single most powerful force in all of Creation—It's not easy to stop order. We believe chaos theory to be disorder, however, it is anything but disorder. Everything reacts upon the next thing and each of those reactions is quantifiable. Our problem is that we fail to see the diversity of possibility of each subsequent reaction.

Two single atoms interacting could have thousands of possible outcomes. Upon interacting the newly formed substance with yet another atom, thousands more outcomes for each possible previous outcome can occur. When we take this intense exponential manner and try to apply it to our "rational" thinking, then our thinking cannot realize the magnitude of the task, thus, our minds ignore the subject after a few generations of exponential growth. This simple truth of order is undeniable and unstoppable.

Two atoms interacting may only have a few potential actual outcomes in reality, but a literal handful of atoms will have what, to us, is an exponentially inconceivable amount of outcomes. Nothing is random, it only appears so to us. Some could take this to mean that we do not possess free will, but this is not at all true.

There is an anomaly in our thinking with regard to free will; and the anomaly is that, if there is a Creator, then free will does not exist. From a scientific perspective, the opposite is true. Free will cannot exist without a Creator because order only allows a finite amount of outcomes for each given situation, where with a Creator *we* get to *choose* outcomes by imagining ideas and then acting on those newly created thoughts

The vastness of order in the Universe is clearly displayed at every level of observation. The Enlightenment movement of the eighteenth century would have had us believe that everything could be explained with Newton's laws. This, then, could be interpreted as rubber-stamping the Universe; but, then, all of the galaxies would be identical.

As we can see, when looking into the cosmos with our telescopes, matter within the Universe is highly ordered and there is zero sign of chaos. We look into our hand and see atoms repeated by numbers so big that debate of this point is futile. When we look into space, we see stars and galaxies in numbers so big that, again, debate of the point is futile.

The order of all matter has a repetition and accuracy so great that humans cannot even comprehend it. We count in tens and assume everything functions in tens, but that is a very short-sighted way to view the diverse simplicity of order. Order does not count in tens. Order may count in numbers that we generally do not grasp, such as infinity, thus giving the illusion that "random" exists. Order simply *is*. The only true chaos that exists is in our heads when we are dishonest with our findings. The Universe works under the power of order, and that order does not shake easily.

When we try to create disorder, then the only thing that will be destroyed is us. Order begets *order* and disorder begets *order*. Anything that is in disorder will either submit to order or it will eventually cease.

We might think an atomic bomb to be chaotic, but it's not, because an atomic bomb does what we anticipate it to do and likely more accurately than we will ever be able to measure without dying in the process; though, we do not know the actual specific magnitude of the released energy, which we are likely inaccurate in our assessment thereof. This means that based upon the order that we see evident through Einstein's famous equation, we have built machines that create a great deal of activity, resulting in sudden heat and expansion. Now that is highly ordered!

What we think of as chaos, is nothing more than an adjustment in order. If we try to create disorder with order, then we will destroy ourselves with the power of order. Order cannot be defeated—it is *everywhere*.

When we send a probe out into space and receive radio signals back from it, then we can detect those signals in various ways; and in this same way, we can listen to the stars and planets. One way to interpret these signals is to slow the recorded frequency to a lower level and turn it into sound. When we do this, we hear what appears to be un-patterned patterns of sounds, or erratic oscillations. Pattern is an inseparable aspect of order that is so diverse that carbon copy duplication will not be recognized or even occur on the upper cosmic scale, yet pattern appears highly common on the lower subatomic scale. Disorder is rare and what we perceive as disorder typically is the corrective nature of order.

All matter is subject to order, and we see this order everywhere we look, in everything we see, and in everything that we do. Even the apparent nothingness that everything is Created from is a form of high order. Even in the disorganized appearance of fractals we see ever-increasing repetition of order as the scale changes. Humanity's lack of, or ignorance of, concrete evidence of the nature of matter allows us to embrace the wrong things or thoughts, and thus ignore the power of true order.

Chapter 9

Has Science Embraced The Wrong Things?

Instead of embracing the fact that we don't understand light and gravity, we ignore that fact and we pretend that we understand. We embrace the bent ruler of our improperly used mathematics and call it *fact* even when nothing could be further from the truth—We have embraced the wrong things!

I am Attracted to Gravity

We simply do not understand gravity, and it would be good for that fact to be better known by the up-and-coming young scientists if we ever hope to truly understand gravity. I am attracted to gravity because gravity is not greatly understood, and is therefore an intriguing field of study that is wide open for new discovery. We have half-baked ideas about the effects of gravity but they are far from actually understanding gravity. It seems fair to say, through observation of the field of science, that there are a good number of scientists that believe gravity is understood by themselves. I say this in reference to and based upon some interviews recorded that contain their own commentary on gravitational physics.

We do not understand if, what we call, "*gravity*" is a foundational phenomenon or if it is an encompassing phenomenon. Meaning, are we on it where it is pulling us, or is gravity somehow pushing us down?

The implications for misunderstanding gravity are many, and we can clearly see the misunderstanding. However, since our scope of usage is fairly limited, we only need to bend the ruler whenever we need to systematically interact with gravity in order to use it to our scientific theory's advantage. We make the needed adjustments and then continue on with our work. From a practical use perspective this is all that matters to us, but since this book is about bending the ruler in the science world, these misunderstandings become very important issues to grasp.

Gravity is said to affect light, and through that we have extrapolated it to bending time and space. We believe that Einstein's simplistic 3-D view of gravity being like a rubber membrane shows us how time and space bend (Figure 7 Rubber-Sheet-Plane, Page 119.) We have accepted this since early in the twentieth century; however, Einstein's youthful illustration does not automatically make altering time or space a possibility.

Personal Relativity

What really is relativity? Can our acceleration actually create gravity or increase our mass? Can our velocity actually slow time for us personally?

It is said that, as our velocity increases, our mass increases and our time slows. It is believed that when traveling, our personal clock slows, yet *we* (the traveler) will perceive it to be normal; but those who we have returned to after our travel will have experienced a greater duration of elapsed time than we will have experienced. Could this be true? If we travel fast are we then younger than our stationary counterparts when we return?

Here is something to think about: With regard to the speed of light, when does light's reference point begin? What happens if you are born into a given speed, then does the light exist at a different speed for the newly-born you? Meaning, if I am not yet born, and my future world is traveling faster than the speed of light, then, will the existing light be in existence equal to my relative speed when I am born? Or will the light remain relative to its initial source? This thought is difficult to convey, as are most anomalies.

Consider this scenario: Let's say that there are two women with husbands. One of the couples will travel very near the speed of light. Now, at the same moment, conception of child occurs in each woman, which is at the moment the traveling couple leaves (for the sake of illustration we will assume the *exact same* moment in time for both women). Also, we will have our hypothetical travelers *instantly* traveling near the speed of light, ignoring the potential ramifications of instant acceleration to that speed.

The two pregnant women are both going to gestate, we'll say for exactly nine months to the moment, relative to each of their own clocks. When the birth of each child occurs, the traveling woman returns home with her family, but the stationary woman's child will be considerably older. In a hypothetical situation I can accept that. But, what would happen if the traveler returned the day the stationary woman gave birth? Would the traveler only be days along in her pregnancy?

Now, instead, let's assume that both women became pregnant *exactly one month* **after** departure according to each her own clock. Now, the traveling woman is moving at a rate that slows time to 1/20 of the progression of the stationary woman's time. If the traveler returns on the day that the stationary woman gives birth, then, will the traveler be impregnated yet? For the stationary woman ten months since departure have elapsed, but for the traveler it is only 1/20 of that duration according to the theoretical relativity concept.

We can even take this mental exercise a bit further: Since the women are contemplating bringing forth new life, we should actually be focusing on the new life rather than on the women.

Going back to our last example of a 1/20 time differential, the stationary woman's child experienced a full accurate nine month gestation at birth, and the traveling woman was gone for ten months according to the stationary woman's child. Will the traveling woman have a child when she returns since she only experienced 1/20 of the ten month trip or about 15 days on her clock? Also, if the traveler thought to return in nine months when the stationary woman gave birth, would she actually return in twenty times the nine months (180 months) for the stationary woman?

Since the stationary child has experienced a standard gestational amount of time, he will be expecting a friend to return who would share the same birthday and age.

The question is this: will the traveling child's conception have taken place at all, since it will have been sandwiched between their—departure and arrival—duration of time in relation to the stationary child? Since the traveling child did not yet exist before departure, when does his or her time start? Does the latent time affect the traveling child's life?

If the traveling child is traveling for nine months according to its own clock, and then heads back to meet his new stationary friend, will it take the traveling child another nine months to return? If so, then does the stationary child continue to age at a 20:1 ratio as he heads back, or is the child's time reversed on the way back?

We are going to assume that the traveling child's parents are going to continue their journey at near the speed of light, but they are going to send their child back at the same rate of speed at which they left, then, can the child ever return? Or will the child be stationary, suspended in space relative to the **stationary** family? Does light change speed for you after you are born? What exactly is the point of *reference* in time travel hypotheses?

Is Time Travel Possible?

Because the mental anomalies with time travel are so many and so diverse, it is difficult to convey them in a few short paragraphs. However, the mental exercises just stated should begin to get people thinking in the right direction. Yes, the examples just given were meant to confuse and raise questions because that is exactly what our juvenile view of relativity does.

Time is not substantive, and it is no different than saying *volume*. *Time* is not measured; it is the idea of the duration of something. Volume is not a measure of something, it is the space occupied in the expanse. Distance is not the measure of something, it is the gap between two points.

Distance = gap between two points location.

Time = gap between two points of duration.

Duration is to *"endure"* and **endure** is to *"harden."* It is fair to say to *"make solid"* or *"tangible,"* meaning: A gap between two points of duration that something was somewhere—or how long it was there. We measure this in days, years, seasons, months, hours, minutes, and seconds. We measure the dimension of distance in miles, kilometers, meters, etc. and we measure volume in gallons, liters, milliliters, etc.

Our mathematics are linear in nature like distance is linear. We can imagine that space is curved, and the distance between the two points is actually not a straight line, but we only perceive it that way in our imaginations—once again, this is bending the ruler. Personally, I am a diehard fan of *"the shortest distance between two points is a *straight* line"*, but with our contemporary, and not so scientific, method of bending space-time, who can know what "*straight*" actually is.

Our ruler has gotten so badly damaged that our worshiping of light has led us to conclude that we are incapable of knowing if a line is actually straight—a thinking which is based upon the belief that space can be bent. This is because our perception of our theoretically bent line appears to us to be straight. With this sort of rationale we have the space of our Universe being contorted into spirals, donuts, folding space and allowing wormholes, and many other obscure forms that have been proposed. But then I ask, what is *outside* of those forms?

Time is not some**thing** that can be traveled or manipulated; it is abstract—it is an idea, it is a measurement. We wrongly believe that *time* is some**thing**. It is true that we can call time a dimension because it is a measure of duration. Instead of thinking that time is something, we should instead realize that *time* is the duration that the thing is *being* there which is being measured using elapsed time with increments of hours, minutes, seconds etc. Existence is not moved forward or backwards, it is just there and we measure how long it is there in the same way that we measure how long, how wide, or how tall something is. Due to the amount of ruler bending that we do while rationalizing our thoughts on light's properties when we discuss traveling through time, our rationale is akin to using a different scale of measure for each unit measured while arranging *length*, *width*, and *height*.

Controlling Time

Due to our interpretation of Einstein's and other peoples' theories, we humans act as if we somehow control "time," yet many of us cannot even seem to get along with our neighbors. If we can't even control ourselves, then how do *we* expect to control time?

The self-centric view we have been discussing is prominent in all aspects of our lives. Few escape the pull of the gravity of a self-centered perspective. Some of the influence in our thinking is done innocently, but much is not. This is especially true whenever increased status is up for grabs. Gaining status causes some otherwise very good people to become stupid and deliberately pervert the sciences, while, with some others, perverting the sciences is done unintentionally.

There are many great scientists out there who have their heads on straight and perceive our surroundings in an accurate and articulate manner. They are always leaving possibility open for contemplation. But there are also many scientists who have bought into the modern blatant myths of certain specifics of time, space, and gravity, yet they claim they are open minded about it.

Time is a dimension, like distance and direction are dimensions. We can control the speed of our clocks and the increments used to measure time, but we cannot alter duration. Duration is measured with "time" and it is continuously and infinitely elapsing forward at one constant rate.

Creating Disorder

Matter is highly ordered, and that which is made of matter is also highly ordered. Each new level of Creation has a magnitude of order that is inconceivable to our minds. According to the theory of long-age evolution, a few hundred million years from now our heads will become enormous in order to contain our massively evolved brains.

In our view of the celestial bodies, we believe that the forces of physics are what order everything. Many want to believe that, at some point, it was all chaos. However, if at any point, pattern arose, then chaos could not exist. Chaos is as imagined in our minds as time-travel is.

Chaos, as stated a couple of sections back, is only our perception of the complexities of what we do not understand or comprehend. In truth, we cannot create disorder or chaos in the physical realm, we can only *try* to create disorder. Order will have its way with us, and it will quickly restore any disruptions we attempt to cause. These corrections reoccur; meaning that the corrections are ordered. This means that chaos and disorder do not exist in nature. The only way we can create disorder is by our choice to do irrational actions that aren't true. When we do so, then order will have its way with us, and we will eventually be destroyed because we have chosen to defy the undefeatable nature of order.

Chance of the Universe

"Chance" of the Universe is not possible because the Universe is highly ordered. The big bang is one of the single most absurd proposals to be made in all of recorded history. Believing that some god created it all, is far more believable than a "big bang" as big bang is proposed.

What exactly is the appeal of some of these absurd scientific beliefs? Why do we insist that it happened from a big bang? Could there have been a big bang for every galaxy? Or for every star? That would be far more convincing. If we're going to get hypothetical, then let's at least make sure that our math works in ninety percent of the situations rather than in none of the situations.

It seems that the more outlandish the hypothesis, then the more attention the imaginative person gets. I have wondered for many years if the people that propose some of these absurd hypotheses actually believe them, or if they are only after the spotlight and money. And after observing this for several decades, it becomes very clear that the spotlight and money means more to many of those scientists, who are the attention-seeking types, than does the truth and accuracy.

We can create mathematical equations "proving" that the big bang would have to eventually occur; but, it seems that the point of it needing to occur depends upon the imaginative capacity of the scientist proposing the new hypothesis. In the end, the "big bang *theory*" is irrational and "random"—it is chance.

I do not believe in chance in this regard. Many may take this to indicate that we can predict everything, but that is absurd. If you can fathom what infinite is, then there are infinite particles creating infinite celestial bodies in infinite Space. This leaves an infinite-infinites quantity of possibilities that all share the same effects of order.

We can anticipate a small scope for just about anything, but when we get too far removed from the original source of the event, then the variability has grown to such an extent that the possibilities are inconceivable to us. Yet, within it all, this apparent "chance" is highly ordered.

All too often we want absolutes and exact answers, but the kind of absolutes that we want are going to be very hard to come by when using the wrong information with which to arrive at our hypotheses. Further, if we had it our way, and our math was absolute and exact, then we would all be identical mathematical clones of one another. All of the stars are the same, yet they are different, and so too it is with humans; we are all the same, but we are very different. Such diversity is wonderful, but it is highly ordered; and all of the stars share this same order—where there is order, "chance" is not.

Chapter 10

What Are We Doing?

We, as a science community, are committing to beliefs that have not been proven. We know that, what we believe to be, anomalies *do* appear in our observations. These "anomalies", however, are only our interpretation of the observation when weighed against our expectations. The expectations that we have are based upon our mathematic equations and upon our interpretations of our prior observations and day-to-day real-life observations.

Anytime we see, what we believe to be, an anomaly occur, or a discrepancy between our math and our actual observations, or between a logical expected result and an actual result, then our best bet is to re-evaluate our prior observations *and* our math.

Be Careful with e=mc²

Be careful with $e=mc^2$ because it is a truly volatile equation that is based upon a great amount of uncertainty. It seems evident that there is a correlation between energy and mass, which is not in question at this point in the book.

Ole Christensen Rømer was an astronomer who made the first known quantitative measurements of the speed of light. It is said that the speed of light has been approximated ever since Rømer had the insightful observations about Jupiter's moons in the sixteen hundreds. This was amazing considering he had little in the way of equipment. His observational method was genius given his era, and interestingly enough, all of the prior calculations that his estimate was built upon were reasonably accurate enough to give him his approximation of the speed of light.

Rømer's logic was that the speed at which light traveled was *finite*, as opposed to the belief that light was instant. His discovery of proof of light's finite speed was a wonderful revelation to humanity. It should be noted that Rømer was wrong in his actual measurement of light's speed. His projected speed of light needed a fair amount of adjustment in order to be close to light's relative speed as we know it today, but his observation that light's speed was finite, as opposed to being instant, was a very important discovery.

Einstein's contribution of realizing an equivalency between energy and mass was another wonderful revelation. But $e=mc^2$ needs a similar adjustment as Rømer's discovery, and that adjustment again comes down to the speed of light. The problem with Einstein's equation is that it has the potential for exponential errors depending upon the way you view and use the equation, and it also depends upon *who* is viewing the equation. For science to continue using these formulas and the somewhat more accurate, but still unknown, properties of light, is simply inaccurate science. These formulas and the unknown properties of light will result in errors when dealing with very large numbers.

Our Perception of Distance

Our perception of how far away a light emitting body appears to be is based upon our understanding of what we believe the speed of light to be, coupled with red-shift. With light, we have a tendency to want to have things multiple ways, so we bend time saying that space-time can change. It's highly likely that neither time (Existence) nor space (Expanse) can bend, change, reverse, shrink, or stretch.

There's a big difference between space and time *actually* being bent, and space and time <u>appearing</u> to be bent; and in this case only *mathematically* appearing to be bent. Our perception of gravitation and light are intertwined. When our calculations on one are wrong, then our calculations on the other will be wrong as well.

If light is variable, then we can throw out Hubble's red-shift idea for gauging the distance of celestial bodies (which is based upon light having a "constant" speed.) Also, if light can actually be drawn into a black hole, then that is another reason to throw out Hubble's red-shift method of dating the Universe and gauging distance of celestial bodies. Even $e=mc^2$ itself shows the potential error of Hubble's method. There is also the issue of calibration of the method: To what extent has the red-shift occurred? Is it gradual, or does it suddenly shift to red? Who gets to set the index for the value of degree of shift in relation to actual distance?

When the measurement referred to as "foot" came about there were discrepancies involved due to the lack of a single standard. Once the idea became accepted, then a standard needed to be designated. Once a standard was designated, the people had a solid reference point for a "foot" and it became a physically tangible stick that could be reused. But discrepancies still existed because the standards were only regional. Because there is little room for obscurity when holding a stick in hand, this standardized foot system worked well. In later years, an

Bending The Ruler

effort was made to internationally standardize measurement, and the result of that effort was the *meter*, which was admirable but was needlessly based upon a fraction of the size of the Earth's circumference. Each time the measuring stick was used it could be physically witnessed. But, when it comes to indices, such as red-shift and the speed of light, then things are a bit different.

A light-year is a confusing term, in a sense, because we are using time to designate distance; a distance which is merely an *idea* rather than an actual tangible wooden measuring stick. We have no way of knowing exactly how long that *idea* stick is in reality because we can only calculate it mathematically. As far as we can tell, there are no celestial bodies large enough where we can lay a one-light-year-long stick out on the ground and walk its distance. The intangible nature of the idea of a light-year is susceptible to tremendous error, especially when that idea could be moving at very fast speeds.

Our perception of distance on Earth is fairly good in relation to our binocular vision and the perspective that we experience with our vision. But binocular perception is lost in the expanse because the distances are so tremendous.

In the end, for our day-to-day life, knowing the speed of light has little importance, we will live and die whether or not we know and understand the speed of light. Science, on the other hand, is an altogether different story: In science, light's speed is critical to a multitude of equations; and if light is variable or our estimate of light's speed is wrong, then all equations using light's speed inherit those errors.

Bending the Rulers

By bending rulers we can force our conclusions to *appear* to work. Those conclusions can even be useful and wrong at the same time because the conclusion, or scientific index, may work well for a finite area of study. Forcing $e=mc^2$ to work by means of bending space-time is a gross error. This stops us from being able to accurately explain science and our understanding of the vast Universe of Creation.

When we take a small section of any graph, then our formula used to calculate that specific graph section will work well anywhere within that section (Figure 3 Stock Chart, Page 70.) We can even project the equation out far beyond the given section, and then write some algorithm to compensate for the apparent anomalies we see. This sort of science has been serving us well, but that does not make it correct. In truth, the best we can claim when it comes to the speed of light is that it is a good approximation. The motion aspects of the light-speed discussion are not *provable* with our current scientific bases.

Accountability of Measurement

When Hubble first proposed the red-shift method for measuring celestial distance, what exactly was used as for the yard stick of index? Of course that sounds ridiculous, but that's exactly the point. He set the standard for the unit of measure, or the *method* of measure in this case. Who is accountable for that standard? Who can show us how long this standard actually is? Hubble may indeed be correct about red-shift, and two equally redshifted bodies might truly be of similar distance from us. However, what that distance is is what is being brought to question here.

Science has peer review for accountability, but that's another story in itself. Accountability is a balance—and *accurate* balance reveals truth. If there is no accountability, then we cannot assume any truth whatsoever. Someone could stand and be heard, saying, "I will accept accountability", but then to what end? Since we cannot place a light-year in a place of safe-keeping in a secured government office, we have no means of proving its consistency; and thus, accountability is meaningless in that case because there is nothing to be accountable for. This allows for a great deal of wild speculation without full tangible proof, for nothing more than financial gain and ego building of the proposing scientist.

The assumptions about light are being taught as if they are perfect fact, much the way long-age evolution is taught as fact even though they claim otherwise. Science has captured the hearts and minds of the western world, and science worships light even more stringently than the Church worshiped God in centuries past. In the Bible, there are several hundred rules and ten main ones; however, with science, there is no end to the rules and laws. If any scientist does not bow before the scientific laws they will be cast out of their community—It seems that history does repeat itself, because decenters of doctrine were excommunicated.

Objects Let us Down

This may appear a little off topic, but since this book is about bending the ruler of science, and since it is *us people* who bend that ruler, a bit about ourselves is in order.

Objects let us down because we want to put our faith in those objects. Take money for instance, it is a great thing to have in order to make life more pleasant at times, but those who attempt to derive their happiness from money often suffer a disappointing life, even though they may be rich. This is because we expect *objects* to become the joy we are truly searching for.

Being let down is why people become angry at objects; for instance, kicking something when you're mad at it. Our problem with objects is that the objects are not supposed to command us, but rather we are supposed to command the objects. Our misunderstanding about our environment has dealt us a tremendous set of blows over the thousands of years of recorded human history.

We are let down because we do not understand. In general, we function well with our erred mathematics because we are functioning within a very finite realm, and, for all practical purposes, that is sufficient.

Humanity has made many wonderful advancements built upon *incorrect* conclusions about some very important scientific points. The finite scientific realm we work in has allowed us to do so, and it may seem that we are never disappointed, but that's not true. How many people would enjoy a brief trip into space? Anyone who is not afraid to do so would jump at the chance to take a brief ride to the nearest star, but our current beliefs are aiding in barring us from doing so because we believe certain things about the practicality of traveling faster than the speed of light.

Our let-down from the objects around us is hidden by our blindness. We cannot exactly call it disappointing when we choose to not allow ourselves to have the dream to begin with, but not having the dream is the biggest let-down of all. The objects let us down because we let ourselves down believing that certain phenomenon has power over us. Sadly, we worship light; and because we do, we have set the object of light as our limit.

I Love Creation!

Whether it was Created by a Creator, or the less likely big bang, or anywhere in between, kudos to that which Created! I love the Creation and its Creator deeply. Every part of men is a part of Creation, and Creation is from a Creator. If that be some random cosmic blunder, then I say well done random cosmic blunder for the artfulness of the male and his female counterpart in the cosmos; that is a job well done!

Wherever we look we see immense beauty; be it in bed next to you or in the expanse in which we dwell, the beauty is immeasurable! The deeper in space we look, the more beautiful it seems to be, and that's only considering the detectable aspects we are currently aware of.

In the same way that we turn the invisible radiation into a visible picture by using different equipment and different spectrum, we can also turn radiation into sound, and those sounds can be very beautiful as well.

The twentieth century and early twenty-first century citizens were given a wonderful gift in the NASA Hubble telescope. Even if there are inaccuracies in its namesake Edwin Hubble's red-shift method for gauging distance and movement, the composite picture of the various radiations are amazing to see. And coupled together with the internet and high speed communications, the citizens of the world have the opportunity to see what the ancients could not imagine. Our Universe is a truly beautiful Creation regardless of how it came to be.

Chapter 11

Receive the Evidence

As we observe the heavens we behold many wonderful sights. The same can be said for the microscopic subatomic realm. For the most part, we observe these sights and accept them for what they are. It's when they get beyond our reach that we start guessing.

Our math is not subject to our sensory perception. Just as we can create mental mathematical multidimensional arrays with computers, we can also imagine other mental anomalies that don't compute in the real world.

Sensory Perception

For the most part, our perceptions of what we experience are based upon our five senses. What we perceive in our immediate surroundings is truly all that matters to us at any given point in time. Whether it is the air or a meal, or anything else, we see it, we feel it, we taste it, we smell it, and we hear it; and, from this, we humans have deduced a multitude of basic perceptions.

What is amazing about our senses is that we are able to extract data from actual experience and translate that into thought for speculative projection. Our senses allow us to imagine based upon what we have experienced in the past. We have a truly creative power about us—we can *wonder!*

When we project too far out without having adequate experience or adequate thought, then we are highly subject to error. Imagine what would happen if light suddenly went the speed of sound; the confusion that would be going on in our heads would be intense.

Our perception of our surroundings is dependent upon a relative amount of constancy. For instance, light is anything but constant and Einstein knew this, but due to light's peculiar and immediate nature (Light does not appreciably suffer from the issues that accompany accelerating mass), it appears as constant from the observer's view. In Einstein's theory, the "observer" is the one who is moving in harmony with the light's source. It is because of this behavior of light that we will likely never experience a noticeable shift in light's speed. The instantaneous behavior of light and its low susceptibility to gravitation are all a part of our perception of light.

In reality, we actually see things *after* the light is emitted or reflected, but the latency is too small for us to detect. So in the end, for practical use application, light seems to be the beat-all substance for us to use to be able to see with.

Does it Matter?

Does it matter that we experience the objects around us slightly after they existed for that moment at that location? No, just because it takes time for light to get to us, does not mean that it matters in any substantial way. Humanity has been surviving quite happily for thousands years of recorded history with no noticeable negative effects from the minor latency of light or the even more prominent latency of sound.

The only time that the speed of light becomes an issue is when we are measuring large values based upon light's speed. For instance, the additive values of the mass of an object affects our perception of how we believe that the object will interact with a gravitational field. The vast perceived distances spoken of throughout this book are also greatly altered by our perception of the speed of light *and* by the perception that that perception has on our perception of time.

I am trying to approach this subject of light's constancy in every way I can think of because it is so error prone and grossly overlooked in the world of science and, most notably, in astrophysics specifically. We have allowed a great deal of speculation to be perpetrated onto society due to the errors in our assumptions about the behavior of light. It is highly unlikely that a big bang occurred in the manner as proposed in the twentieth and twenty-first centuries. It is also highly unlikely that it occurred 13.7 billion years ago *if* the big bang actually happened at all. The big bang and red-shift dating are two beliefs that need each other to survive. One without the other breaks down the method of calculation. Another factor that goes hand in hand with big bang and red-shift dating is light's perceived constancy. These three are akin to some of the crazy inaccurate *speculations* about Biblical prophecy over the years, and they are all one in the same type of arrogant error.

Time in a Black Hole

As mentioned earlier, a "black hole" is a theoretical concept. It is very important to understand that a "black hole" is a poor choice of name for the phenomenon. A *black star* or a **dark** *star* designating "*hide*" or "*hidden*" would be a far better choice of terminology. The idea that a hole in space exists, is insinuated by the term "*black hole*." It is understandable why someone would use the term *black hole* to describe a black hole, but the term "*black hole*" is better suited for sci-fi than it is sci-ence. "*Black holes*" likely exist, but are inappropriately named when using the term for scientific purposes. Viewing a **black star** as if it is a *black hole* is a fantasy based perspective, rather than a perspective based in reality.

Relativity extrapolated to the magnitude of a massive gravity star (black hole) allows for the bending of space and mind rather than the bending of space and time. Taken far enough, this bending can be bent to an infinitely small size as is demonstrated in the big bang singularity hypothesis. Our evidence of matter actually shrinking to that extent is left extremely wanting. Taking our math beyond its original scope is, in this case, a very foolish error.

I do not believe that black *holes* exist, however I propose that a black star or a dark star can easily exist. When we properly understand light and its interaction with matter, it is not a stretch to realize that light is pulled towards the star and cannot escape the surrounding gravity of the star. This is commonly referred to as the *event horizon*, which is another term that is a very narrow view of the scope of matter.

A question to be asked is, does light *suddenly* shift from radiating out, in an instant, to being drawn in, similar to flipping a light switch? Or does it change slowly and progressively until it finally reverses direction?

When light is overcome by gravity does time slow, stop, or reverse? Since "time" is only an idea, we should be able to realize that the idea of time cannot change, just as the idea of distance cannot change. While atomic compression can take place to some extent, and can change the physical dimension of an item, it is unlikely ever to altogether collapse to zero point size. Similarly, light's speed is likely to change, and our theoretical perception of time changing will be based upon our assumptions about the correlation between light, time, and the constancy of light.

A black star will not allow light to escape its gravitation, meaning that any light particles or light radiation passing nearby will likely be drawn into the star, but the black star need not be of any peculiar unrealistic mathematical density.

Why does certain radiation get drawn back to the Sun as witnessed during solar flares? What are the dark spots on the Sun? Are the Sun spots evidence of the Sun becoming a black star? Will *our* Sun become a black star? Is our Sun turning into a "black hole"?

Is our Sun Turning into a Black Hole?

Because science does not have an absolutely clear understanding of gravity, we fail to deeply grasp the nature of light and the reasons for, and possibility of, black holes.

So, could our Sun be on the verge of becoming a black hole? And if it ever did, would it consume everything near it? No. The Sun's spots have some very indicative patterns when viewing them with a powerful specialized sun-telescope. The density of the Sun is unknown and could vary from area to area of the Sun's surface. Light is likely variable in speed, which means that it would be affected by gravity in different ways depending upon the force of the gravity of its emission source area. If light is susceptible to gravity, and since visible light has a narrow area of spectrum, the ability of gravity to affect light would have a very small range of gravitational force that would need to change for light to stop radiating outward.

The Sun could be in a balanced state and the Sun spots may likely be a part of that balance. If the Sun didn't ever release energy, then it is likely that it would already be a black hole. But since solar flares and energy emission are constant, the Sun is in balance and is not likely to become a black hole. Yet the Sunspots do seem to indicate increased gravitational force. As for the planets being sucked into the black-hole-Sun, it would not necessarily happen that way because the needed increase for gravitational force to stop light from being emitted may be very small if the gravity is already near the threshold of being unable to emit light. So, while Earth would not be pulled into the Sun, we would potentially lose our visible sunlight and possibly our heat.

Size of the Universe

Infinite is **infinite**, and if a person cannot wrap their head around that concept, then science is probably not their best occupational choice. The size of the *Universe* has been "*infinite*" for a long time, meaning it is not finite. Intuitively, this means that the Universe is not limited; and thus, it has no borders and no end.

In the late twentieth century, a one hundred-year-old theory re-emerged that claimed multiple universes exist. This is a somewhat short-sighted hypothesis because the Universe is either infinite or it is not. If we are going to choose to decide that it is not infinite, then we should make a declaration on the terminology and keep that index firm and change the name accordingly. Stating that we should keep the index firm might appear to be in conflict with the message of this book, but it is not; it's quite the opposite. We must define our idea of "*Universe*" and stick with it or more of our mental walls will be allowed to be repeatedly set up at the edge of each larger encompassing "multiverse" that we imagine within our minds.

Our gauge of the speed of light, the size of the Universe, the distance of stars, and age of the Universe are all tangled up in a few potential mathematical errors and oversights, allowing us to believe things that might not be true. Science has become a religion whose god is light's constant speed, and whose prophets are the likes of Einstein, Darwin, and Hubble.

Our Ability to Think

Our views and perceptions can only be as open as our ability to think freely, so the ways in which we see or describe time in our mind is limited by each our own minds. We have already discussed *time* at length and what time is, and it is critical to understand and remember those points. So to repeat: **time** is an *idea* that is used to measure *duration* in increments of seconds, minutes, hours, etc.

When we bind ourselves to our religion, we become beholden to any lies and inaccuracies that may reside within its doctrine, or within our own interpretation of that doctrine. At the beginning of the twenty-first century, science was in the same state that the Church was in several hundred years prior to that time, but science had a different god. From my observations, depending upon the individual interviewed within the science astrophysics community, their understanding of Biblical issues was more accurate (whether or not they believed the Bible) than their scientific beliefs at that time of the beginning of the twenty-first century. Science has itself caught in a very dangerous state of being—Setting our indices up as our idols only traps us deeper in our error and inhibits our ability to think freely. Then we additionally deepen the stance of our incorrect position through our arrogance and through our fear of admitting our errors.

Scientific Testing

Our inhibited thinking has led us to believe that our scientific testing is "proof" of accuracy, just like the Bible became the "proof" for the Church leaders' beliefs.

It is important to understand the following: An experiment using flawed data is flawed throughout; thus, the testing is inaccurate.

Any results produced from the experiment will exhibit remnants of the flawed data, but to us they will appear correct because we have blindly accepted the flawed index as true. This is the same as when light is said to appear to be going the speed of light to the moving observer. The perception of the observer who is moving with the pack of scientists will observe the tests to be accurate, but to someone with a different point of reference, that pack of scientists, no matter how many there are, will appear to be going in the wrong scientific direction.

Just like many people blindly believing the **preachers' interpretation** of Genesis chapter one (that the entire Creation was made in six 24 hours days) is the only way Creation could have occurred, so, too, is believing in many of our scientific laws and theories regarding big bang and evolution as the way things are, without having solid proof. This point must be stressed because the problem is in an epidemic state in the science world, and it will continue to be so until someone speaks out. While the language in the schools states the case with wording that sounds open for reinterpretation, the truth is that few teachers, and few people in the field of science, have an attitude to match those words—and their "laws" support that inaccuracy. For instance, something is often called a "theory" when it is nothing more than a mere hypothesis; yet, as we discussed earlier, it is addressed as factual as if it has been proven with hard evidence.

When we do scientific testing, we experiment with various atoms and compounds, then we see results and we analyze those results in our minds. We run tests for everything that we can think of in order to understand what we see. Our tests on the elements and all that is made from those elements are not actually tests on the elements—they are tests of our own minds to be able to see the truth with regard to how the elements function. Adjusting our perspective to see it in this way, and understanding that we are testing *ourselves*, and not the chemicals or physical phenomena, will bring science to a whole new level of understanding.

I feel confident in saying that the order by which the elements abide is stable enough that the experiments that we do today will have very similar results as the experiments did in the past, and as the results will also likely be in the future. This intuitive observation indicates that the elements do not need testing because they always do the same thing; however, our minds do not always come to the same conclusions about what we believe we witness the elements do. But even in this, we lack long-term experience, for instance, with regard to elements degrading. If they do degrade, then at what rate do they do so, and is that rate constant, or is it affected in any way by other nearby elements or gravity?

As is so clearly illustrated by history, two minds can look at the same thing and see two entirely different possibilities, and usually only one or neither of them is accurate and true. Though, if they appear to differ, it is possible that both are true depending upon what each person is seeing and how they described it.

We must realize that we are testing our own minds, rather than imagining that we are testing the elements or other physical phenomena. As far as we can tell, the elements and other physical phenomena have remained the same through hundreds of years of testing. When we realize that we are testing ourselves, then we begin to realize that we are trying to see what *is*. Through understanding this, we will try making fewer commandments, so that we can begin to make additional, and more accurate, descriptions of what we perceive that we are seeing.

Chapter 12

The Ruler

What is a **ruler**? A ruler is something that rules of course! Which brings us to the question, what does **rule** mean? **Rule** is to "*keep straight, direct, or right.*" It's safe to infer that when we refer to a measuring instrument by the term **ruler**, we are speaking of a direct straight line between two points represented by the incremented stick.

In our minds, a common desk *ruler* is usually a plastic, metal, or wooden stick that has various increments on it to represent arbitrary spans of space. Who sets these arbitrary spans? And does it matter? As discussed earlier, a light-year is also a ruler that we use, but since we have no means to actually lay down, produce, or use a ruler of such immense length, we are left with finding a mere representation for a ruler with that immense of size.

A light-year takes our arbitrarily marked desk-ruler and mentally assigns an enormous amount of desk-rulers (or other measuring devices) in an imaginary end to end manner to measure out the amount of distance light would travel in a year, but this is a terrible misnomer.

Is a Constant what We Say it is?

Since science is believed to be very particular in its methods, so it seems only fair to apply the *foundation* of those methods to the same rigor. The foundation of any area of study is the means of communication used within that area of study. In science, we believe the primary means of communication to be math, but it is not. Math is just as well suited for stirring up a pitcher of lemonade as it is in a scientist's lab experiments. The primary and therefore foundational means of communication for science is the *words* used in working with the math, for instance: 1+1=2. If we cannot agree upon the value of the words used to communicate the ideas represented in our math, then we can clearly infer that the value of all of the math that is used is equally in question, if not more in question than the words themselves are.

Our words are primary to **everything**. And if we cannot come to a consensus on the value of those words, then we have greatly hindered our own understanding of our observations and the sharing of those observations with others. Some might argue that science does have a scientific language, and that the language is very specific, but that is invitingly debatable. Additionally, the science index-words are not only subject to the specific definition that we choose to arbitrarily assign to them, but they are also subject to the interpretation that *each individual scientist* assigns to them.

Let's take the word **constant** for instance, **con-stant** is to "*com-stand*" or "*co-stand*" or "*com-stare,*" it is to be "*firm.*" A "*constant*" is an idea of unchangingness. But in truth, a human defined constant is always relative to something that is less constant. This means that if we find something more constant than our old standard, then our old standard is no longer "constant" by definition. Einstein's view was that light's speed does not change from the perspective of the person traveling along with the source of the light. This may well be true, but since it is mostly not provable, it may also be very wrong.

A constant is what *we* say it is in our own personal scientific reality, which is very dangerous to the advancement of science. We intuitively understand what is meant by the concept of "*constant*," but this does not mean that—what we assign the value of *constant* to—is actually *constant*. Light as a "*constant*" is constant only in our minds, and we have every indication that this is so **when** we read the data in a particular way. The reason that light is allowed to stand as a scientific constant is because the need has not yet arisen to be more constant than light for life's general sustenance. For practical purposes, as far as we can tell $e=m$, and the c^2 part is irrelevant. If the situation ever occurs that higher accuracy is required, be assured that it will not take long for all of society to come to the realization of light's non-constant nature.

Our opinion on the constancy of a "constant" is what *we* say it is, but this does not mean that light is actually "constant." "Constant" is relative to the most constant thing we are able to detect. Any *constant* other than the order found in truth is arrogant to conclude. One thing that is certainly constant in the science world is our ability to misunderstand and misinterpret what we see and what we think we see occur.

In the Presence of Gravity

Space-time is said to slow when near to intense gravity, but the perception of space-time slowing is because of our assessment of light's interaction with energy and mass. We see things only as we propose them until someone has the courage to thwart the system. For fear of losing their job or status, few scientists will dare to turn against the consensual culture in science and proceed to debate any long-held erred belief.

Consider entertaining the idea that light is not constant in any situation, and that light's speed can be manipulated. What are the scientific ramifications of that? In truth, there will be little effect on our day to day lives when this is finally accepted. Just because light's instantaneousness inhibits our ability to detect its variability, does not change the way light actually behaves. So whether we are right or wrong about light, life will still go on as usual. Light is anything but constant. Believing that it is constant will only serve to hold us in an even more Aristotelian-style of scientific dark ages. As for substances, light is the most stable substance known to us. That level of stability is admirable, but that does not make it a "constant."

We can make adjustments to light similar to what is done when building a structure upon a crooked foundation. The problem with building on a crooked foundation is in the adjustments placed in each layer's bricks within the foundation and building. When constructing a building on a crooked foundation, we would use extra mortar to compensate for the foundation errors, but the errors will still exist; and the errors will to some extent, affect the final product. In working with light, we build on the crooked foundation that forces us to make "adjustments" or calculations in order to allow us to believe that the experiment is mathematically straight, even though it appears crooked through our physical observations. Thus we make yet more adjustments to compensate for the persistent error. When you have error in your calculations, then the errors will remain inherent in all subsequent results from those calculations.

Newton's "laws of physics" are similar in effect to our assumption that light is constant. When we take Newton's "laws" and apply them too far outside of their observational parameters, then we have no real evidence that they are still fully useful. We can say that Newton's laws "break down", but then we must ask: Were they intended for infinite use? If they "break down" then are they really "Laws"?

Again we can revisit the example used earlier of a child growing: Can we actually take the average growth rate of a twelve-year-old child and extrapolate that to their eighty-forth birthday? We all understand this to be absurd, but we lose this fundamental scientific reasoning when it comes to physics because we have what we refer to as "laws". Most of those laws are not provable when working with immense distances and extremely high velocities.

The "Laws" of Physics are a Lie

At the beginning of the twenty-first century, anyone found to be in violation of the long held "laws" of science was ostracized by their peers and also by the scientific and secular media. The humiliation for pointing out the truth that opposes consensus carries a heavy price and can ruin a career if a scientist says the wrong things. This is readily observed when researching that topic, which I will leave for you to do. Since credit scores dominated western culture early in the twenty-first century, few scientists dared to turn against the consensus of culture in science for fear of losing their job or status, which could result in leaving them in the poor house. This sort of situation can only occur when our beliefs have become our religion and we succumb to being rated; meaning that we are going to believe what we believe because that is what we learned from our superiors, **and** we are afraid to thwart that because of the potential social and financial ramifications we will face. Many scientists do not want to hear this because they have dedicated their entire lives to proving their erred hypotheses, which are based upon a bent ruler that they chose to believe was straight.

There are no such things as "laws" of physics; there is only order and *our* explanation of that order. Coming to this understanding sooner rather than later will bring science to a new understanding about what *is*. What can we take away from this? Well, that's up to your ability to adequately reason through what you see; but please consider that "laws" are manmade, and they're only man's feeble attempt to make sense of what we believe we experience, and it is no different for the laws of the land.

Using the term "law" in science is not a very scientific approach, and it is in direct conflict with the scientific method. What we call the "laws" of physics are nothing more than mathematical descriptions of our observations within the scope of our ability to conduct an experiment. In truth, our laws can only apply accurately to the particular scope from which they were taken. Using the "laws" for anything outside of the original scope is an extrapolation of theory, and the margin for error increases with each further deviation from the source point experimentation.

Does Disorder Exist?

We discussed chaos earlier and we touched on *disorder*. We think of *chaos* similar to *disorder*, or even the same as *disorder*, but is it the same? This, of course, will depend upon your interpretation of the two terms *chaos* and *disorder*. **Chaos** connects to the word "*abyss*" which indicates "*bottomless*." Does this mean that *chaos* is *infinite*? The word **disorder** indicates "*apart from*" or "*two*" and then together with "*order*." Depending upon your interpretation, *chaos* and *disorder* could mean very different things.

Order is primary in all of Creation. It seems that there is little that defies order—other than our free will to do so. Everything else follows order with perfect precision as is demonstrated by all that is seen in the expanse and by all that is seen at the atomic and subatomic levels. Order is powerful and takes its time, but it never ceases doing its work. When we try to create disorder, then order immediately begins working to restore order. But, in reality, order never ceased because order was *always* present. Even when a nuclear device is detonated, order still prevails, and all of the activity occurs as anticipated by our limited expectations of order.

True disorder is the utter and complete cessation of order, and that cannot ever be. If there is no order then we have nothing tangible. If something comes into true disorder it will immediately cease to be; disorder will come into order, or it will cease to exist. If this were not true, we would not be able to predict *anything* in science. Our scientific predictions are based solely upon our very loosely made observations of order's stunning reliability. This is true with our day to day lives as well; such as the reliability of the ground that we walk upon, or the reliability of the air that we breathe.

The Universe is not 13.7 Billion-Years-Old

To recap, the most distant galaxy cannot be 13.2 billion years old (it is believed by some that it took a half billion years to form) because the light that is here today would have been emitted 13.2 billion years ago. This means that the distant galaxy as being viewed today is at least 13.2 billion *more* light-years away and is 13.2 billion years *older* today if that distant part of the Universe has not yet dissolved or contracted. This assumption is based upon the big bang theory, but the big bang also has various sub-theories, and depending upon who is discussing the theories, the Universe expanded at many times the speed of light. The big bang concept, which uses Newton's laws, is very shortsighted and has many anomalies, and even more so when we assume that the speed of light is constant.

According to twentieth century calculations, for light to be reaching us today from 13.2 billion light-years away, the source object would have to have been in that location 13.2 billion years ago. This means that the most distant galaxy known to us moved into its current position in 0.5 billion years or less. This is a 26:1 ratio. This means that, at a minimum, if the most distant galaxy, and our galaxy, started halfway in between, we then had to have moved at thirteen times the speed of light, each moving in a direction away from each other for 0.5 billion years. If light travels at the speed of light as prescribed by Einstein's "constant", this means that it took 13.2 billion years for the light to reach us. If the light emitted from the most distant known galaxy did emit the currently seen light 13.2 billion years ago, then that galaxy is actually now 343 billion or more light-years away from us, assuming the travel speed of us and the distant galaxy remained constant; which, of course, is adhering to an ever-expanding-universe based theory, but there are also theories that state the expansion is slowing, and some claim it is contracting now, and yet others claim it to be static.

According to e=mc^2 we believe nothing can exceed the speed of light, but the big bang says otherwise. Further, according to Einstein's Theory of Relativity, the source moving away from us is going to appear to be moving at a slower rate, as per Einstein's "Twin Paradox" (which we will discuss later.) We believe we have verified the twin paradox mathematically. So, does this mean that the movement of us and the most distant galaxy cancels each other out?

Let's say that this movement does cancel. So instead, let's make Earth stationary and move the most distant galaxy away from us at twenty-six times the speed of light. Now, in this scenario, does the light actually ever reach us? Is the galaxy going back in time because it is going much faster than the speed of light? Since the speed of light is the index point, then exceeding this index point must surpass light's speed, and according to the scientific indexes and Einstein's relativity, at a minimum, this slows time to a standstill—time is frozen!

Using the constant speed of light indicates that the most distant galaxy had to have been in position no less than 0.5 billion years after the initial moment of the bang in order for us to see the light as we do today. The furthest galaxy had to have traveled faster than the speed of light at a constant rate of no less than twenty-six times the speed of light. However, since the galaxy had to start from zero speed, then, at some point, it had to have been traveling much faster than twenty-six times light's speed in order to get to its current perceived location in enough time for us to see it there today.

We can experiment with gravitational forces to try to explain away the speeds if we want, but in every situation we must violate our own "laws" of physics in order to force the Universe into compliance with our juvenile imaginings. It's clear that there are gross errors in our "laws" and/or in the application and perception of those laws. If time is slowed with motion, then what exactly is our perception of that distant galaxy?

Projecting Time

We want to believe many things in multiple ways. We want to claim that there is no God, but that aliens planted us here. Or that there is no Heaven, but that there are other dimensions that are within our own space area where we currently reside. It's very unscientific reasoning for us to say with certainty that something does or does not exist without having substantial evidence.

Have I trapped myself in those words? No, not at all. Our reasoning for our denial of things is based solely upon each of our own chosen set of beliefs. Remember, that to *believe* light to be constant is a *belief*; it is a religion and light is its god; when in reality all evidence points to light being variable.

Einstein's vision of riding a light beam could be thought of in terms of a movie projector (Figure 8 Movie Projector Timeline, Page 176.) If you ran the projector for a few minutes and then began to move at the speed of the projected light and in harmony with the projected beam, then you could index each movie frame's image. Since you are still moving in reference to the light and at the same speed as the projector's light emission, you are not seeing any movement of the movie frames. If you were to begin to advance in the light beam by exceeding the speed of the projector's light beam, then you would see each frame regressively pass in the index of movie frames. This would cause the projected movie to run in reverse for you the observer. Based upon big bang theory, and due to the speeds and distances that would had to have been achieved with big bang (which are due to the logic we use), our view of the furthest galaxy is not its age at 13.2 billion years, but rather, it is how it looked billions of years *before it existed.*

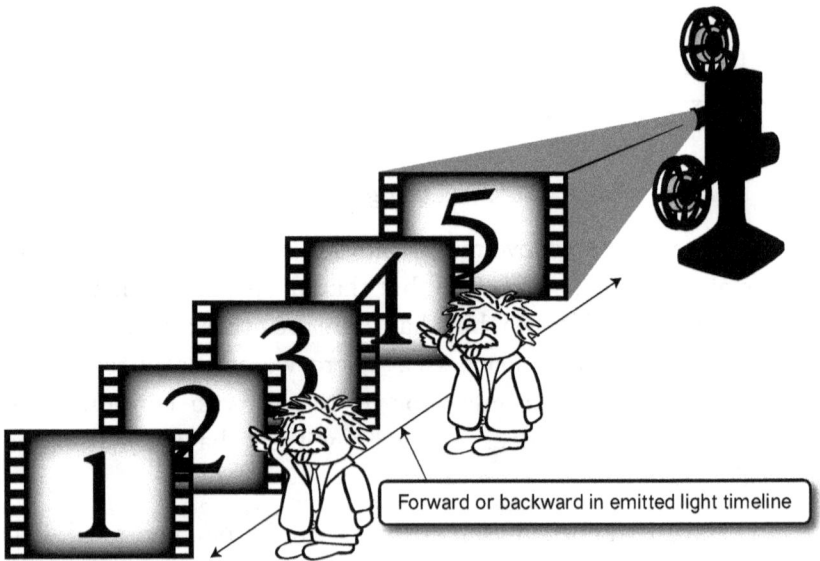

Forward or backward in emitted light timeline

Figure 8 Movie Projector Timeline

In big-bang-theory terms, the distant galaxy is possibly being seen today as it was in a previous bang cycle. This could mean that the big bang has collapsed and expanded, and what we are seeing today is light that occurred before the big bang. Oh, but the many anomalies in that idea... This could mean that the galaxy is collapsing towards us, but at what speed we do not know. However, this presents us with a problem called red-shift. If the most distant galaxy is moving towards us, then what happens to red-shift?

You better be prepared, because if the Universe can collapse as fast as it theoretically expanded according to big bang theory, then at any moment we may be crushed into oblivion and we won't see it coming because it is moving faster than the speed of light. We will not see it coming because we are only now seeing the light that occurred 13.2 billion years ago; meaning that the distant galaxy could be at our front door according to the rules of the big bang theory.

We have mangled the scientific ruler so badly when it comes to "*black holes*," the "*big bang*," the "*constant*" speed of light, and the "*laws*" of physics that we could not find our way out of a black hole even if there was an alternate dimension and universe on the other side of it.

Chapter 13

The Ocean of Our Universe

From our best guess, the Universe is a vast ocean of moving matter that is in constant motion. Assuming that the laws of physics are remotely similar throughout the entire Universe, we can make a safe assumption that everything is in motion to some extent. Since *size* is a relative physical aspect, any celestial body is relatively no different in size in the infinite expanse than a single molecule of water is an entire ocean. This vast repetitious ocean of our Universe is there for our enjoyment, and we should observe it and attempt to understand its infinite and endless value and beauty.

Multiple "Singularities"

The big bang assumes that "singularity" is real, and that **all** matter was in singularity at some time in the past, meaning that all matter was compressed to a single point of an infinitely small size. And from that singularity, came forth all matter; but is this realistic? When asking about the validity of singularity, I am not referring to whether or not a *finite* amount of matter can be drawn into a single location of singularity, but rather, I am asking, can an *infinite* amount of matter to be drawn into singularity?

Why is it so important to so many scientists for all matter to have been "big banged" from a single point? Since, as best as we can tell, the expanse is infinite, then couldn't there be infinite big bangs that are constantly occurring all over the expanse? The obvious answer to this is, "yes, that could be a possibility", since infinite means without any borders and without end.

The singularity of the big bang is the single most narrow and short-sighted approach that we have ever heard with regard to understanding the origins of the Universe. Singularity need not have happened at a single point; "Singularity" could have happened simultaneously at an infinite amount of points, thus creating an infinite amount of matter.

As proposed here, multiple "singularities" are far more likely to have been infinitely occurring throughout the expanse of time and space creating galaxies with each different singularity, rather than entire universes coming from a single singularity. While it is likely closer to reality, even multiple singularities is somewhat short-sighted in regard to the origins of the cosmos.

Here again, it comes down to definition: If the singularity of the big bang created the Universe, and we had infinite big bangs occurring, then are the infinite universes in one big multiverse? Where do we stop this bending of the *definition* ruler? The Universe, as proposed, is infinite and therefore there is only one **Uni**-verse. We can play mental games and we can do the mental theoretical exercises, but that won't change the fact that the term "universe" inherently unendingly encompasses **all**. Since the term universe was, at one time, thought to encompass all, then why do we need the term multiverse which implies multiple universes? This literally means multiple *all-encompassings*.

Assuming that a *multiverse* exists is a copout for our minds and allows us to set up barriers to infinity when we can't answer questions and conflicts that arise in our minds due to our misinterpretation of the "Laws". If we allow these barriers in our mind, then eventually we will only come up with an additional theory that there are ***multi***-multiverses and ***multi-multi***-multiverses.

Depending upon the age that we assign to the Universe, our perception of the reality of time itself is greatly altered. We believe the Universe to be expanding, but outside of the Universe-dating system that we call "red-shift", all signs point to an attraction of matter, rather than an expansion. So we can trash our "laws" of physics in regard to scientific big bang consensus.

Ask yourself this: Is it likely that the Universe is cyclical? Does the big bang occur again and again, each time collapsing and expanding?

The Speed of the Galaxy

According to the current consensus in the scientific community at the time this was written, the most distant galaxy is 13.2 billion light-years away from us; and it is speculated that all of creation is 13.7 billion years old. Based upon this consensus of understanding, the speculation is that everything came from a single point source since the big bang. The most distant galaxy had to travel at speeds far beyond the speed of light in order to move far enough away from us in time for us to be able to see the light from that galaxy's currently perceived location.

There is another problem with the big bang: These distances and speeds blindly assume linear dispersion, completely overlooking radial dispersion problems, illustrated in the following diagram.

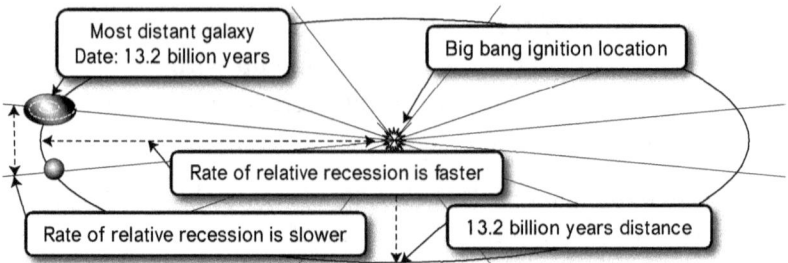

Figure 9 Radial Universe Dispersion

Radial dispersion will vary greatly depending upon the degrees of rotation of the two observation points. Based upon Red-Shift methods, any degree of deviation from 180 degrees will greatly affect our perception of the speed and distance traveled when we include the consideration of radial dispersion. To expand with a separation distance parting at the speed of light at a given departure angle, would cause the radial dispersion speed to be exponentially faster with any closing angle of a greater degree.

The point of discussing this is to illustrate that relative to the position of the source ignition point of the big bang, our short-sighted view of the big bang very likely has us moving faster than the speed of light at this very moment. Using all conventional math and logic, based upon the big bang theory, our galaxy is moving well beyond the speed of light. There is no reference point as to what direction we are looking or facing in reference to the big bang. The other celestial bodies that we see may well be moving near to parallel to us with regard to the radial nature of a big bang explosion.

If the big bang truly did occur, then we should see tremendous difference in our observations depending upon which direction we look relative to the big bang's source point; unless, of course, we place ourselves and our Earth at the very source point of the bang. As far as can be determined by our red-shift data, the big bang could not have occurred as specified in science. Sadly, it seems that science has once again gone Earth-centric in its understanding of the Heavens.

Do Black Holes Really Exist?

So as to not leave any thought behind, we will revisit black holes and the constant speed of light. Since "black holes" have come to be widely accepted in the science world, we must scrutinize them with scientific vigor. However, for our purposes we will stay close to the ruler-bending aspects of the topic.

Light is said to be constant from the source perspective. So according to this it means that if I am moving in perfect harmony with the emission's source, my perspective of that light is constant. When a "black hole" emits light out into space, does that light suddenly drop back or is the effect of gravity on the light beam gradual? To state this better: If I am in zero motion relative to a star, and a light beam is being directed into space perpendicular to the plain that I and the light emission source are on, then if the gravity of the star increases to a point where the light can no longer defy the gravity, does the light slow down or does it suddenly turn off as if it were switched off, and then fall to the star?

If light is constant, then it would have to either be on or off; thus eliminating the possibility of an "event horizon." (Even the originator of the idea of an "event horizon" finally realized that his theory was greatly flawed.) If a black hole can attract or absorb light, then the speed of light is likely not constant. Additionally, the angular perception of the light's speed is going to be different based upon the principles laid down by relativity, as shown in the following illustration. The perception of motion is different for each perpendicular increment the more distant you are from the plane of motion.

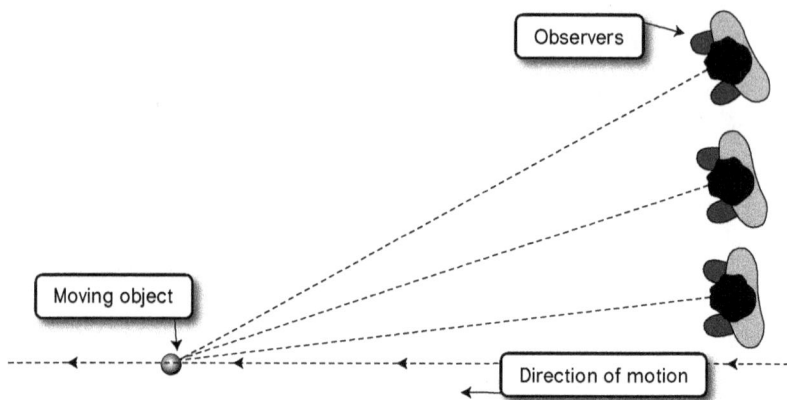

Figure 10 Angular Light Speed Perception

A black hole is a terrible misnomer that relies entirely upon the idea of singularity. If singularity does not exist, then there are no black "holes." As stated earlier, what we believe to be a black hole is more appropriately and paradoxically named a *dark star* or black star. A dark star is a reasonable name for a phenomenon that we believe we have observed in the Universe. Demanding that a dark star must collapse into singularity at some point is likely defying reality.

The terminology "black hole" is best left to the over extension of our finite mathematic laws. We can play with the term black hole for fantasy movies and experiment with mental exercises regarding the perceived paradoxes; however, for practical and progressive science, we are far better served to refer to high-mass stars as black or dark stars, or simply high mass stars. By this we are indicating that they are still stars, but with a power of gravity that has enough strength that even light will not escape its gravitational pull. We cannot see a dark star with conventional methods that we normally use to see the light of most stars. However, other stars that are orbiting around the area where we believe a dark star to be is reasonably strong evidence that such a phenomenon might actually occur, but we must consider that two stars can orbit each other, making them appear to be orbiting a central point. It may very well be that there are dark high-mass stars nearer to us than the nearest visible star is, but we cannot easily detect them because they do not emit any visible light for us to detect.

Does Mass Increase with Speed?

As discussed earlier, our determination of our cosmic speed is going to vary greatly depending upon our assumption of our trajectory from the source point of the big bang. If a singularity big bang did occur, then our speed must presently be greater than the speed of light. There can be little debate about this without grossly violating every "law" of physics that is deemed absolute by the majority of the scientific community in order for big bang to have occurred.

Mass is believed to increase with velocity, but does it actually do so? We believe that we prove this in particle-accelerators; however, there are some oversights in firmly holding this belief. All of our scientific views are theoretical views backed by finite experiments conducted in a finite space. What would the mass-gain-differential be in a particle accelerator on the moon for instance, or floating in space far out of the reach of any substantial gravitational field?

We believe that the mass of an accelerated particle increases with increased velocity; and it may well do so, but this speculative thinking is currently difficult to measure, and so, it cannot be adequately proven. The mass changes that we perceive can be attributed to other factors. Our insistence of utilizing Einstein's handy perspective and its derivations has served us well, but doing so so rigidly has inhibited our future knowledge.

We may very well be able to accelerate a particle far beyond the speed of light, but that is unlikely to ever occur with our current scientific mindset because when we do these experiments we are trying to exceed a human imposed "speed limit" that likely does not even exist. Electromagnetically pushing a particle around in a particle accelerator limits the acceleration to the abilities of the electricity and the fields that it creates, even when it is alternately timed in cycle for faster acceleration.

We command particles when we turn on a five dollar flashlight. Our multi-billion dollar approaches may be spending billions to do what could be done for only a few million if we were to try to understand things in a *true* manner—Although... particle accelerators are very cool toys!

Waves of Good Vibrations

Waves and vibrations go hand in hand. The metaphysical sector has taken the vibration thing to a whole new level and has set vibrations (the oscillation of matter) up as the god of metaphysics, just as light has been set up as the god of physics. But, we should not discount the value of either *light* or *vibrations*.

Both *light* and *vibrations* are phenomena that serve us well on a regular basis. Again, it comes down to our understanding of these two phenomena with which we deceive ourselves. Waves and vibration are not fully understood, though it does appear that we do have a good grasp of them; yet, our mental vision is often lacking regarding them. Vibrations seem to permeate all of Creation, and while these vibrations are good and useful, they are not magical as is often proposed in metaphysics.

We have this blind-eyed view in our world that the magician works magic even when we actually know better, and metaphysics is no different: In metaphysics the spirit realm is confused with the tangible world. It is often believed that some sort of cosmic vibration is the carrier of all good things. All of Creation is stuff, and that stuff is all around us and contains no magical powers. However, we can manipulate it for our good and for our curiosity, all for the betterment of mankind. Regardless of the source of matter, we have all been given this wondrous gift which we should use properly.

Since light "waves" are not fully understood, our continued contemplation in that particular branch of science will serve us well. In science there seems to be an unspoken view that light travels in beams. For instance, when we point a laser in a particular direction it shines a beam of light out in that direction, but we need to think of light a bit differently when discussing its emission. Light energy is typically dispersed in a radial orb-like manner from the source. If I am seeing the energy from a single atom, does a person 90 degrees from me, as measured from the source-atom, see the *same* "light wave", or can only one of us see that particular wave? This question is connected to the quantum physics duality view of a particle being in two places at one time, or in neither place, or in either place, as is taught in science classroom to many unsuspecting school children. Our view on this is very narrow, but we have enough information about light's peculiar behavior that we can do many amazing and wonderful tasks with it, such as using it to measure local distance or speed where the inherent error is irrelevant. Sometimes we even use it to the point that it becomes dangerous to us.

We intuitively grasp what a vibration is, but our view of a vibration is very two dimensional. This is because, in our minds, we typically have a picture of the waves traveling on a single plane that warps as the wave ripples through the plane. This water-like view of a wave does science more harm than it does it good, but since it is a reasonable representation of the outward motion of a wave it is regularly used as an illustration of light waves. Obtaining a better understanding of "vibrations" and "wave function" can bring science to a new and more advanced position. Deviation in color intensity is a far better visual.

The Oceans of Europa

This chapter, *The Ocean of Our Universe*, has the purpose of pointing out the fluidity of the cosmos and *our* inconsistency in interpreting what we see. We even believe that the moon of Jupiter, which we call Europa, has an icy layer nearly 100 kilometers thick, and liquid water is beneath the surface. This is interesting since the evidence for water on the surface is speculative.

Europa is not an issue, nor are the speculations about it. What is an issue is our inconsistent understanding of what we see around us. We should take issue when we can say that Europa has water on or in it that is believed to be double the volume of the water in Earth's oceans; yet, we cannot conceive any possibility that a cataclysmic flood could ever have encompassed the entire Earth. This statement could ruffle a few evolutionary feathers; yet, we cannot explain the far too many anomalies encountered when justifying early twenty-first century theories in geology and so we choose to ignore them.

This is not a call to blindly follow church-views on six-day Creation, but we should re-evaluate our take on the structure of the Earth and what happened to put fossilized fish in some highly unlikely places. We should also seek a unified theory of the Earth's origin that does not contradict itself or ignore *any* evidence.

Just as our interpretation of the church's interpretation of Genesis One has errors, so too, does our own interpretation of science's interpretation of what we have observed also have errors. Will we believe that all matter was compressed into an infinitely small point? Will we believe that Europa possesses water that is double the volume of the Earth's water, but that Earth can have no possible hidden water? Will we believe that life came here from Mars in space debris, but could not have begun here?

Chapter 14

Is There Life Out There?

Our new religion of science, combined with our misguided worship of the speed of light, have dimmed our ability to see newer and greater things, just as the church leaders' view dimmed the eyes of the people from seeing newer and greater things over past centuries.

We Earth people are self-centered. We were self-centered back in the pre-Copernican days, and we are self-centered in any contemporary time. And now, it seems that we have gone back to an Earth-centric view of creation in using our big bang theology, and, once again, we are limiting our ability to see. Even if we deny it in our words, our big bang view places us near the radial center of the specific ignition point in most mathematical models. When we accept flawed theory, we are left without understanding and we begin, in our quest to explain our origins, to believe things such as, our life here on Earth hitched a ride from Mars, but we fail to then explain how life arrived and survived on the very barren Mars.

Is There Other Water Out There? Is There Life Out There?

The evaluation of water existing anywhere in space is dependent upon our estimation of what we are actually seeing. Are stars truly similar to our Sun? Are those spiral things that we see truly galaxies? Do we actually live in one of those spiral things similar to what we see out in space?

If we are correct in our assumptions of the stars, of the galaxies, and of our own galaxy, then the probability that there is good clean water on other celestial bodies anywhere in space is very high. The problem that we face is, once again, the same self-centered view: We insist that for "life" to exist there must be water; and/or that if there is water, then life must exist. I do not debate either possibility, but I do debate our narrow view that all life *must* involve "water." Our definition of life seems somewhat bound to water because it is what we personally have experienced—it's all that we know. It very well may be that life requires water, but this particular perspective about water lacks the same over-zealous creativity that was used to derive the big bang, and is therefore inconsistent in its scientific methodology.

In reference to the Bible, nowhere in the Bible does it say that there is not life beyond Earth. We humans have an attitude as if the Earth revolves around each of us individually. Just as we did with the Bible for so long, so too, we have also done with science: Due to our misdirected interpretations, we believe things that our written accounts of evidence do not specifically state.

The problem with religion was not the alleged "fables" in the Bible; the problem with religion was the same as it is now with much of science—it is a personal perspective viewpoint. With the Bible, many assumed that *they* were being told the **whole** story from the Creator, which is a somewhat arrogant perspective. We are doing the same today with science: We believe that we know the **whole** story about gravity, light, the speed of light, and so much more. But we do not know the *whole* story; in fact, we have barely just begun!

Flawed Science

Science has some errors and/or potential flaws in its thinking. For instance, why are meteors often said to be from Mars? Some scientists believe that surface fragments from Mars flew out into space upon an asteroid impact with the planet, and then eventually these ejected fragments landed on Earth, bringing primordial life along with it inside of the rock itself. And further, many also believe that this is how life began on Earth. This short-sighted view fails to answer how life arrived on Mars.

Could it be that the meteors found on Earth, which are thought to have originated from Mars with what is thought to be possible evidence of life, are actually from a planet other than Mars? And maybe even from Earth itself? Could an asteroid have landed in a primordial Earth ocean and blasted Earth rocks into space that subsequently are caught in an orbit around the Sun, and then while those fragments are traveling near to the Earth, at some point the Earth passes through the primordial rock debris trail?

If the energies released from a large asteroid hitting Mars are enough to eject Mars rocks into space, then these energies are certainly enough to eject Earth debris into space that will maintain a speed different from Earth's speed in orbit around the Sun. It is then possible for that hypothetical impact to cause the debris to orbit for thousands of years before returning to Earth, giving us the erred illusion that it came from Mars. If these asteroids could come from Mars, then they could certainly come from Earth as well. We truly cannot say what the composition of the original material was that may have been ejected during a meteor strike of that magnitude. Debris could theoretically be ejected that originally lay far below the Earth's surface—further below than mankind has yet been able to achieve and subsequently analyze. The heat and energy from an asteroid impact would alter the structure of the asteroid and its debris upon entering the atmosphere and upon impact, which would be very misleading to us. Further, the debris would have been ejected and then would have re-entered the Earth's atmosphere in a fiery ball and impacted with the surface of the Earth. Why do we keep returning to the belief that life had to start from a source other than Earth? Is it possible? Sure, it's possible. Is it likely? Probably *not.*

As of the writing of this text, from the evidence that we see and from the documentation that we have, including the Bible, neither science nor the church can definitively confirm or deny whether or not other life exists in the heavens. It is arrogant to insist upon a position of cosmic life and that it *must* be one way or the other based on such information.

If what we see in space is as we believe it to be, then it is possible that there are countless more habitable planets out there. And given the stringent manner of order, it is likely that other life would closely resemble what we see here on Earth. Just like atoms are all similar and the stars are all similar, it is likely that the in-between (from the smallest atom to the largest celestial body) will also be similar. Of course this is using the same extrapolation techniques as are used in science with the laws of physics and many other scientific "laws." But here I call only upon your fundamental ability of logic.

Born at the Speed of Light

Utilizing contemporary models of travel, light speed, and $e=mc^2$, is it possible to travel interstellar and still be alive when the traveler reaches the destination star's solar system? In other words, if intelligent life does exist, then can that other life actually make it here and be alive to talk about it? Will they exist when they get here? How old will they be upon their arrival?

We touched on this subject earlier. What happens if you are born at the speed of light? This question is somewhat related to Einstein's twin paradox. In the twin paradox, a set of twins are going to separate, but only one of them will travel from the starting location. Since they each are their own reference point with regard to motion, they would both appear to age faster than their twin who appeared to be moving away. Since we believe that time is relative to the source, we believe that time for the moving source is slowed. Whose movement is the paradox in reference to? How do we prove who is moving?

Of course, these are only mind exercises, but these ideas about the transformation of time are treated much differently than mere ideas or mere mind exercises. These ideas are treated by many as "fact", but they are not fact; they are mathematical theory and they do not necessarily have real life practical applicability.

Here is another perspective to contemplate: If light is traveling and a mother is observing that light while traveling the speed of that light in the same direction as the light beam is directed, then, in theory, does it change her perception of the speed of that light? If it does, and her world is traveling at the speed of light, and then suddenly you are born or conceived while she is traveling, then is it only at that point that you would be moving relative to anything? Since everything was already in motion, do you see the light as perceived as if you are stationary in relation to the light? Or do you see the light as perceived when moving as fast as your new world when you are conceived or born?

Our view on the speed of light is very short-sighted, and we will gain nothing by continuing to believe wrongly about light, about what it does, and about how it behaves, especially regarding interstellar travel.

Arbitrary Time

If we did not measure time in the way that we do, then we would have a different view of time. This would be similar to the confusion people have when they are taught the metric system, relative to the conventional system of feet and inches, or vice versa.

Time is extremely arbitrary; not so much the idea of time itself, but the actual measurement with time. Placing an index for time is not an easy thing to do. What would a person use for a reference point for time? *Time measurements* are also quite arbitrary.

Our 24-hour day is based upon a single rotation of the Earth in relation to the Sun divided by 24, and then those divisions are each divided into 60 equal parts and so on. This is our best reference to being able to count time. If the motion of Earth and all else in the cosmos could suddenly stop, then our perception of time would be lost within "days."

The invention of a **day** (as we typically interpret it) was a brilliant way to keep time. To have something rotate at a predictable rate and then orbit around the Sun at a very constant rate for an extended duration is nothing short of brilliant! All of our time references are based upon this very brilliant mechanism. We adjust our clocks to the rotation of the Earth around its own axis. This dependable reference has been reliably used for thousands of years with regard to time keeping. We have means of counting and indexing time that are possibly more finely divided than Earth's position, but that is irrelevant and does us little good independent of the Earth's rotation.

If you were to be sent into outer space and had no contact with anyone, other than yourself and those with you, then your time keeping device would become your index until it broke. At that point you would need to find another index, which would likely be to look out the window. But since you would be moving in one constant straight direction, your index would be constantly moving and thus constantly changing for you. And you cannot use the stars as your index because they will not appear in the same relative position as seen here on Earth. Nor would you have the constant motion of earth to index with.

If you did have a very accurate clock on a trip to outer space, then how would you know that it's accurate? The accurate clock could slowly change its calibration and you wouldn't be able to detect that it was changing until it got noticeably fast or slow relative to your at-rest breathing or heartbeat rate.

What we cannot know is the constancy of time measurement. Our measurement system may be affected by many factors. Some say the tidal action of the Moon slows the rotation of the Earth. Some even go so far as to believe that the friction of the winds on the mountains of Earth slows the rotation of the Earth. The thought of winds slowing the Earth's rotation seems a bit naive because, in relation to space and the Earth, the atmosphere is a part of the Earth and moves with it. But the moon's tidal action could conceivably pulse Earth's rotation slightly, but not change its long-term rotational speed.

We believe our atomic clocks to be the most accurate to date, but are they? The value of the atomic clock is not in its accuracy over long periods; the value of an atomic clock is in its decimal point precision; that is to say in the fineness of its divisions. It is able to time events to incredibly small increments; but is there a problem with an atomic clock? Is an atom affected by gravity and speed? It seems that it is. This means that atomic clocks must be calibrated for their velocity and location so that they stay synchronized with Earth-based clocks; so in a sense, atomic clocks are not really that accurate at all, rather they simply have very high resolution.

But what about electricity? Does Electro-Magnetic Interference (EMI) and other radiation affect an electron's speed and atomic activity? Can an atomic clock be altered from, perhaps, a solar flare or other massive bursts of radiation? We cannot say for sure whether an atomic clock does or does not specifically change slightly from these things. If *particles* are at all affected by various cosmic radiation, then all atomic clocks and even standard electric clocks could be affected, and, at minimum, clocks could potentially change equally, or not be perceivably different from each other. Our best clock ever, is the rotation and revolution of Earth.

Since *we* orbit the Earth (being on the outside surface) or are a part of it, Earth is our reference point. We have found no alternate means to readily live outside of the Earth, so Earth's movement is our best reference. Earth's rotation is a standard that is so constant, that in our individual lifespan we will likely see no appreciable change in rotation. Also, since, from our perspective, the Sun passes overhead in a cyclical manner, everyone on Earth can use this same reliable night and day index. Without the index of the rotation of the Earth, our lives would be very different today.

All of our perceptions of speed, light, time, space, and life are all based upon our own experiences. We are only able to imagine what we allow ourselves to imagine. Even if we can imagine well outside of reality, that does not make our imagining true. We can imagine all sorts of life in the cosmos and that there are many other Earth-like planets that we could live on. These are all possibilities; however, one thing is certain, and it is that we will probably never know due to our current outdated arrogant scientific methods.

Aliens

Is there life out there? Who knows? Based upon the logic used to derive this view, the twentieth century view that alien life exists is pushing the boundaries of reason, but who is to say that aliens do not exist?

The idea of other life in the Universe is not nearly as imaginative as asserting that life hitched a ride to Earth on a rock from the planet Mars. I'm not saying that this could not have occurred, but it is a bit sci-fi in nature. First, life would have had to have developed on Mars, and then made it to Earth inside of a rock. This rock would have been cast out from the distant planet's surface while suffering enormous heat and impact-shock. The heat from the energy and impact shock during asteroid collision would likely have cooked and crushed any signs of life. Secondly, the Mars rock had to make its way millions of miles through the extreme radiation and cold of space, and then enter Earth's atmosphere without completely burning up. The rock would have undergone tremendous heat during the entry into Earth's atmosphere, and any chance of life would mostly have been removed. The rock would also have had to once again withstand the enormous impact-shock, both, while entering the atmosphere and then again upon colliding with Earth itself not to mention the added heat produced due to the actual impact.

A Mars rock would have gone through at least two intense impacts and at least one, if not two, immersions in heat so intense that the entire rock was likely glowing red hot or hotter. The likelihood that any life would have survived those events is limited to close to zero. This doubly complicates the theory of evolution in this case, because not only would the evolution of the microbe had to have occurred on Mars, but the unlikely or rare event of a rock from Mars actually making it to Earth, intact, with life surviving, leaves us with the problems of life having to actually escape from its encapsulating rock tomb, which has a probability of slim to none. Beyond all that, the single lucky alien amoeba would have to evolve *and not die* in the process of evolving. And finally, billions of years of evolution would have to take place from that single lonely amoeba in order to evolve all of what we have here on Earth today.

Life originating from Mars is far more imaginative than Genesis or even Darwinian evolution. Adding chance upon chance in this manner creates a magnitude of low probability that can barely be conceived, and even our eloquent numbers and scientific notation are stretched beyond reason in doing so. When we trap ourselves in these naive theories, we cheat ourselves out of what can be.

Based on current science, to insist that either there *cannot* be other life in the Universe, or to insist that there *must* be other life in the Universe is not provable. However, to realize that other life **could be** is the best choice for us right now because we still lack definitive evidence one way or the other. Any other cemented view will blind us because the cemented views are what we insist on promoting to serve our own agendas. If the Church is honest in its interpretation of the Bible, then it must be admitted that nowhere in the Bible does the Creator say that only earth humans exist in the entire Universe, or that there is not life elsewhere. However, the Bible also does *not* say that other life does exist. From a Creator's perspective, I would guess that it is none of our business, since we humans cannot even survive a century without major bloodshed amongst ourselves. What would happen if we could travel to other worlds? Would we do as we do here on earth with our arrogance, robbing, killing, and plundering? An all-knowing and wise Creator would understand that we must first master ourselves before we have any need to experience other worlds.

Insisting that because the cosmos appears infinite does not mean that life exists; and further, *insisting* that life does exist will not make it so. Leaving cosmic extraterrestrial life open to speculation and possibility is the best way to approach the topic. We must keep our eyes open for evidence and we should not try to prove anything; but rather, we must always attempt to see what is true!

Chapter 15

Perceived Structure

What holds all the elements in place? We have our perceived mathematic structures for gravity and light, but these are only descriptions of what we believe is occurring. Yet, the Earth moves silently along its orbit around the Sun far better and more consistent than our clockwork mechanisms or anything else that humans have ever made to track what we refer to as "time".

There are no cables or chains attached in the cosmos; just a sturdy balance of speed and pull to allow a consistent cycle that mankind cannot seem to reproduce, which is made evident when our satellites plummet to Earth's surface in the face of our plans and best intentions. We have divided Earth's consistent cycles into ever smaller increments with our measurements, but to measure and keep time, we still use the consistent stable nature of the cosmos. That consistency becomes our index to notate duration and to calibrate our time increments. In science we pretend that we have the unseen structure of the cosmos all figured out; when in truth, we have barely just begun to understand *any* of it.

The Sands of Math

The general view of mathematics needs a new perspective. In science-calculations, we want to take the math that was developed in order to explain a finite space, and then apply that math to **all** that we see; and in doing so we bend the ruler. Math needs a specific set of rules for each finite area of study.

As an example of this point we can use sand. When sand grains flow through a sieve, only the larger grains are restricted and the smaller grains flow through unaltered. Then those smaller grains can flow through a finer sieve, and again the larger grains will be held back and the smallest grains will flow through the finer sieve.

Each level of sieve has a different set of parameters that apply in order to sort the sands adequately; different rules apply to the smaller grains than to the larger grains.

The Gravity of Math

With math we make the false assumption that things always stay the same. For instance, the pull of gravity is different in space 1000 miles from Earth than it is here on Earth. Saying that light is constant is like saying that the pull of gravity is identical in space to what it is here on Earth. We say that space-time changes when in orbit around Earth; but often, in our thinking, our view is altered in order to fit our imagination—or our lack thereof.

Because we have actually experienced the differences in time-keeping devices during air travel and in satellites, we have been able to make assumed calculations that work adequately in order to adjust for the time-keeping differences that we believe we have witnessed. These calculations may break down if we go outside our current scope. Our math cannot account for this because we do not yet have the needed data due to our current lack of ability to travel at the needed speeds and distances in order to accurately assess light's and gravity's true properties.

The gravity of using our math beyond its practical scope has an attraction so strong that few scientists or theorists escape its pull. The temptation to use our finite experience to calculate the unknown is there because there is little other choice until we actually find a way to experience it. Once we experience it, then we can create mathematical formulas to relay these new experiences to other people.

The rigidity with which we extend our calculations is the issue that we are discussing here. We need to try *something* in order to get the needed information that is required find the truth. However, we should be cautious in our application of this. Insisting that something is a certain way does not make it the way we insist upon; and when we insist to others that it is a certain way, we then teach them to be blind to other possibilities. This is especially adversely affecting young people just entering the field of science.

Our math accounts for what we have seen, but extending it beyond and expecting the same result to infinitely continue is like taking a water soaked sponge and expecting to continuously wring water out of it—infinitely. There is a point at which the water will be removed and the compressed structure of the sponge is all that will remain. Can that structure be compressed to an infinitely small size?

Naming Our Baby

At this point it might sound like the case is being made to have it both ways—imagine, but then don't imagine. This is not the case at all. We should try a crazy speculation like the big bang to see if it fits, and when it does fit then we should explore it even further; however, when the signs point to conflict at *every* turn, then it's time to reconsider our perceived understanding of the theory.

Terms like "big bang" are descriptive; and according to accounts of the origin of the term, it was coined as a mockery of the idea that it represents. Much to the minter's dismay, the term was embraced by the news outlets who love to sensationalize headlines for attention to sell more advertising space, and so in this way the term "big bang" stuck.

Whether accidental or by design, the terms that we use for communicating an idea or phenomenon are there for efficiency of communication. When people spend a great amount of time in research and understanding, and they become the first to realize something, then that is often the person who gets to name his or her baby. It's a very proud moment to have an idea embraced by the scientific community; because, in general, it means that you have been accepted.

The names that we choose to give to our babies, that is to say our projects and theories, and not just in the scientific world, are often somewhat descriptive in as few words as possible. This is in order to convey the general concept without having to re-explain the entire concept with a day's worth of oration every time we discuss the particular topic.

The labels or names that we put on an idea or phenomenon such as *strong force, weak force, gravity*, and *electromagnetic* carry a great deal of meaning with them. Anyone who works in the particular scientific field that encounters one of these terms can efficiently convey a general idea with a single word or a very short phrase. Without these terms or phrases we would not get much accomplished in our world because we would have need of re-explaining the entire concept every time it is discussed.

Strong force, weak force, gravity, and *electromagnetic* are the four forces, or rather the four names that we give in effort to describe the forces that we believe hold all things together. The problem that we face is we have a tendency to accept a stated term as completed, and we act and go on to use it as if it is completed, when, in reality, few if any of these terms are anywhere near complete. Some of the *ideas* about an observed phenomenon that lie behind the names that we give to our scientific babies, are ideas that are only in their infancy. The rest of the ideas that we currently have are not yet even fully gestated.

The Copout

If an atomic clock deviates in space, then it appears evident that the *atoms are affected* and have their activity altered. But, suggesting that space or time is altered because of that clock's time-measurement-change is an entirely different matter.

Changing time or space instead of changing light is a copout. We do not need acceleration or velocity in order to bend space. We can alter the dimensions of an object by merely applying a small amount of heat or removing a small amount of heat to make it expand or contract.

Mathematical extension of equations beyond their initial observed scope causes what we think of as *anomalies* or *paradoxes*. However, the anomalies or paradoxes that we see with time or space are only our mental imaginings, rather than being reality.

It's a copout to ignore and disregard things that do not fit. I believe that there is a unity out there, in the unknown, that we are failing to see. I further speculate that this unity is far simpler than we could ever have imagined. Reality is obscured by *our perception* of reality.

I admire scientists with the spirit and attitude of the likes of Copernicus, Galileo, and Newton who all had a bold approach to look beyond. They also had the ability to reliably prove their hypothesis within the scope of experimentation during the era that they each lived. It is this rare quality that most of the scientists who have taken us to the next level all share.

Increasing Error

The apparent constancy of light is a very practical means of measurement when compared to using an hourglass or some other alternative means of measurement. It might sound as if the text you are reading is saying we should not use light because we cannot rely on its speed, but that is not at all what is being conveyed here. In the end, *measurement* and *index* are all about *scope* and *frame* of reference.

Using a desk ruler to measure a kilometer distance is not the best method to measure a kilometer. Using a meter-long stick or even better a long steel tape would give a much faster and more accurate reading. The smaller the increment, the more inaccurate the end result is likely to be, but is that true? Not exactly. The issue with smaller increments is the margin of error with regard to the length of the measuring device. Marking a kilometer with a one hundred meter steel tape allows for nine hairline errors, where a meter stick allows for one thousand hairline errors, and a typical desk ruler about three thousand hairline errors. When we have a solid structural device with which to measure, then our errors are additive and can be plus or minus as we go, thus, ending in an error that is essentially irrelevant. Light, on the other hand, is different. Our perception of light, mass, and energy has certain exponential error capabilities that become apparent only when carrying the mathematics to extensive proportions. Remember, anything that gets squared increases in an exponential manner as change occurs.

Chapter 16

Interpreting Observations

Interpreting observations is the single most difficult challenge in science. What we see, and what we think we see, are often two conflicting aspects of science. There have been many astute observers throughout the recorded history of mankind, but as articulate as the cumulative observations have been, the *interpretations* of those observations are an entirely different story.

Blinded with Science

When I think of people like Copernicus, Galileo, and Newton, I have a tendency to imagine that they had it easy and did not have to fight the contemporary scientific consensus of their day because there were still so many of the basics to be discovered. They did not have laboratories full of high precision apparatuses with which to conduct their experiments as we have in our contemporary times, yet they were still able to make revolutionary breakthroughs and change the way we all see the world.

The stumbling block that they had to deal with was religious in nature, and the church leaders were at the helm. Today this is different; now we are legally and socially free to deny the Creator and we can freely postulate that everything came to be by some form of natural god-free occurrence without us having to suffer the penalty of our excommunication from the church. But with science this is not true. Many scientists misunderstand light and gravity. Their current belief of what would happen if we exceeded the speed of light is imaginatively speculative. In general, science could care less about today's church. Science, for many people, has replaced the church and is not friendly to those who defy the ways of science. A scientist, so bold as to defy convention or common scientific consensus with good sound logic and demonstration, will often be unjustly discredited and excommunicated from the scientific community just like Copernicus, Galileo, and others were from the church.

It seems that the more ridiculous the claims, then the more attention those claims garner. It's good to be imaginative, but if we are not careful, we will blind ourselves even more with our science and cast ourselves back hundreds of years into the dark ages. This is a type of time travel in which we should not participate!

The Multiverse

Earlier, when touching on the multiverse, we were discussing it from a definition standpoint with regard to the definition of the term "*universe.*" Prior to the late twentieth century's revival of the multiverse theory, I personally understood the Universe to be infinite, and I accepted that as logical. To me the Universe was **all** *encompassing*, but then came the resurgence of a multiverse.

The view of the Universe having been infinite until the multiverse theology came along is a part of the reason that I wrote this book. I am not opposed to participating in these mental exercises with a hypothetical multiverse, but the problem is that pop-science embraces the outlandish, and promotes it as absolute to young unknowing minds, thus, deceiving them; and then, science further goes on to deny the obvious and logical.

Saying that there is more than one Universe shows both an imaginative view and a narrow view at the same time. Depending upon your view of the term multiverse, it can be said to set up walls against infinity within someone's thinking. Are there inter-dimensional multiverses, in other words, multiple multiverses?

Beauty is in the Eye of the Perceiver

The indexes that we create in order to measure and describe all of the beauty that we see around us are only rough approximations of what is actually occurring. Some of the equations built by various scientists over the generations are truly beautiful simple explanations that have served us well; and their equations will continue to serve us within their scope of experimentation.

There is a simple beauty in numbers and equations that cannot be conveyed by words alone. To some people, numbers even have a mystical sense about them. So much can be explained with numbers that, to some people, it appears as if numbers have mystical powers. Believing in Santa Claus does not make the workshop at the North Pole real, and so it is likewise with numbers. Numbers can accurately count what we see because of the pattern and consistency seen within the order that all of Creation is assembled with. Making believe that order obeys our equations is absurd. It's not the intention to rob anyone of their views, but rather the intention is to place science in a reasonable and *realistic* perspective.

Let's continue to see the Universe with the speed of light, but let's calibrate our vision to what is more likely the logical and true outcome.

Can We Actually See Light?

Light is believed to be a substance with particle matter—photons. As far as we have scientifically discerned at this point, light is an excitation of particle matter broadcast in waves. An interesting question is, can a light wave be detected or seen without hitting something and without interference?

What seems to be the peculiar behavior of light is not an easy thing to observe. We currently cannot prove that we have any means of detecting a light *"wave"* in action. All of our observations interfere with the light wave to some extent as we try to detect it. This renders our tests inaccurate at this point in history.

Light is peculiar because if we're in the expanse of space and we're blocked from direct view of a light's source by something that the light hits, then we cannot see the *"beam"* of light by looking from a position approximately perpendicular to the beam of light. Light waves appear to travel invisibly and then become visible upon contact with the receiving entity.

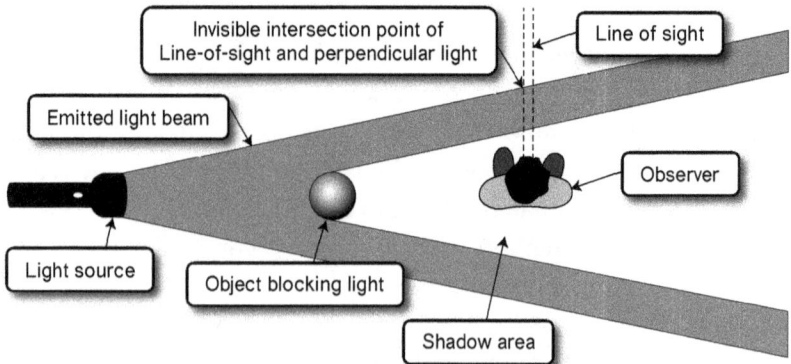

Figure 11 A Perpendicular Light Beam is Not Visible

Our view of light being a *"wave"* would be greatly altered if light did not have the observed apparent interference pattern that is seen in the double-slit experiment (Figure 1 Double-Slit Experiment, Page 50.) Because light seems to remain invisible until it hits what is looking at it or seeing it, we have a very difficult time studying it. As light passes through any sensing-field, the field will likely alter the light's state to some extent. This leaves us in the dark about light until more brilliant experiments are designed and conducted and more accurate hypotheses are made. We are able to detect what we believe is "*frequency*," but as to whether or not we are able to detect an actual "*wave*" is an entirely different story. Light may very well **not** travel in waves, but the double-slit experiment still raises some interesting questions.

What are the ramifications of perpendicular wave interference? Do we have any "*wave*" interference when two light beams cross paths? (Figure 12 Perpendicular Light Interference, Page 209.)

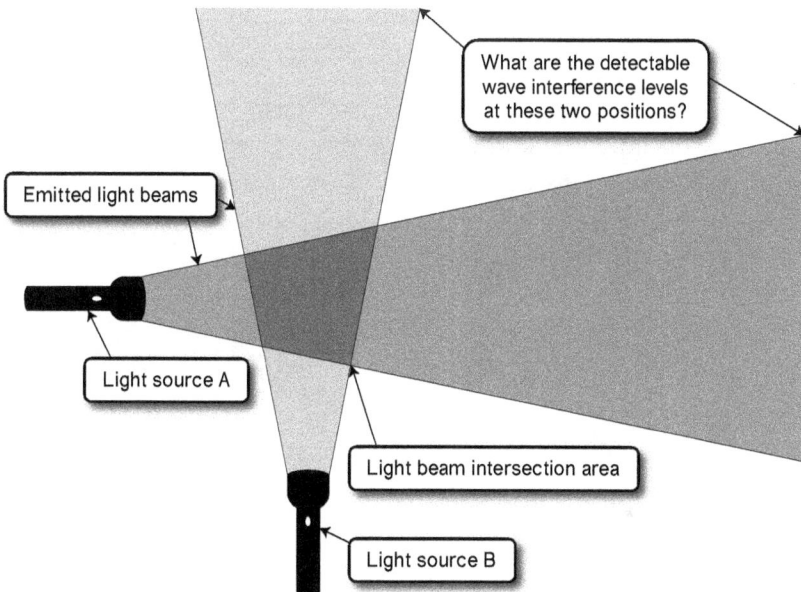

Figure 12 Perpendicular Light Interference

The Perversion of Their Science

Many great scientists were trying to prove the existence of their Creator and better understand the Creation. But, their contributions to this world are often perverted by subsequent scientists. I dare to say, that it is likely that they would not be very pleased with some of the contemporary perversions of their prized theories and laws.

The bold innovative minds of the fathers of science would likely be disgusted with some of the narrow viewpoints with which their work is used in contemporary times. The extending of their work to go beyond the scope of good sense in order to disprove their Creator, for whom many built their hypotheses in order to know better, is not only a perversion of their intention, but is also a slap in their brilliant faces. In their minds, *truth* reigned supreme and they wanted to know *how* **and** *why*, *regardless* of what it meant to their own beliefs. Much of their work was based upon their own understanding of their Creator.

In regard to a Creator, whether you agree with them or not, it must certainly be admitted that some current uses of their work run counter to their desires. These patriarchs of science would certainly frown upon much of the contemporary approach to their scientific methods.

Chapter 17

What We Know

Science is our quest to better understand what we experience. A disconnect from reality occurs when we begin splitting hairs on research topics. In the same way that the Sages split hairs about the meaning of the Torah, so too do scientists split hairs about the meaning of observations of our environment. It is our quest, and it should be our quest, to understand our origins. Science has two key roles in this quest: The first role is the observation of everything, and the second is the dissertation of those observations.

When we feel that we can repeat an observation of experimentation with consistent accuracy, then we share that information with the world to show the grandeur of the Universe and all that it contains. However, we also do it to be recognized for our contribution to science and the world.

We must share what we know with the world so that all people can be aware of the power of the order that we see in the cosmos and below or within. Sadly, many alternate ideas that are accurate, and true, are often overshadowed by the quest to be personally recognized by means of the fantasy theories that have far too many flaws to be realistic or possible, which tend get much more attention than is deserved, for nothing more than mere personal recognition. The personal quest for high scientific-status runs so deep that some people are even willing to steal away credit from others who have truly earned it. Lying about things in order to hide guilt or shame, or to gain notoriety is only a small part of why people do what they do in this world. Certainly not all of science takes part in such poor practices, but these practices do exist—think Nikola Tesla.

Agnostics

Due to the strong stance that some church leaders have taken over the centuries, many people have turned away from the Bible only to insist that it is all fairytales. Some scientists who believe in a Creator will not dare to speak freely of their religious thoughts for fear of being ostracized by their peers. Instead, they only *claim* to be agnostics; where, on the other hand, some scientists actually are agnostics, which when you think about it is technically an ignorant position.

To be a "scientist" and to claim to *not* know something (*agnostic*: "without knowledge") is odd, especially since it is science that claims things such as big bang and evolution to be absolute. Claiming an agnostic belief is an easy way out, but to claim ignorance is better than being wrong.

It's possible that someone can believe a Creator exists and still be open to alternative proof. It also is possible to not believe in a creator and still be open to alternative proof. Though, it seems that either case is rare. Those who are scientists and have adopted the big bang *and* God because of their lack of understanding, have somehow disconnected the Bible and their science. Based upon consensus of the available scientific theories, these two views do not reconcile, and yet they co-exist in the minds of some scientists.

It's peculiar that someone would deny their creed's source book because of equally questionable science. I tend to hold to the belief that it is not our observations that suffer, but rather our interpretation of those observations whether they are scientific or Biblical. For some reason, many people believe that there is no harmony between science and the Bible; or rather, between science and the Creator, but I differ in this: If there is a Creator, then, to me, that is all the more reason to expand our mental scope and to try to see all that is out there.

Provocative Headlines

Generally, exploiters always have been, and always will be the leaders of everything; it may not be fair, but it's true. Exploiters are not necessarily bad, it's just that they understand how to take an idea and get publicity with it, and then capitalize on it for fame and/or for fortune. Again, consider Tesla and the AC current versus DC current issue.

Tesla struck a deal with Thomas Edison that did not work out to Tesla's advantage. Eventually he sold his idea and patent for rotating fields to Westinghouse, as it is told, for a nominal fee for which he was never fully reimbursed. The man who gave us AC current died alone and broke. Yet, all of us use his contributions every day. Thank you Mr. Tesla!

To the credit of the promoter of AC current who gained enormous wealth from it, AC current *is* here today. Often in such cases of money, power, and prestige, the exploiters rule. Those who have taken the power of control will often inhibit those who are the true creators of an idea; these powerful exploiters have taken ideas and called them their own, outright stealing them— thus, the need for a patent office. But even the patent office often fails true innovators when the true innovators eventually succumb to the expense of litigation invoked by people with very deep pockets.

Exploiters and promoters are not shy, and they are willing to take risks and to be controversial. Making a well-timed announcement that both shocks and outrages society will gain a promoter a great deal of attention, and they know this. When the term big bang was mentioned it only took one headline to spark the fire that ignited the big bang, and it was launched to infinity in the minds of many believers.

News thrives on provocative headlines; without shock value the news would lose a great deal of its followers. The more shocking the headlines are, then the more it seems that more people will pay attention.

Attention = Sales

Sales = Advertisers

Advertisers = Money

In general this is a very good system, but sadly it's often abused. The media can help some people get their message to the world, but the media is now, was in the past, and will be in the future, far more concerned about glitz, glory, outrage, and the subsequent money, than it is concerned about the truth.

Much of the fanfare that comes with provocative headlines is damaging in the long run, but it pays big dividends for the promoter or exploiter in the immediate time frame. *They* gain, while the rest of us suffer from their exploitations and, sometimes, from outright lies as well. This should not be taken to mean that all science headlines are a rouse, but there is a disproportional amount of questionable hypotheses getting an unbalanced amount of time and attention from the major media outlets.

What Caused the Big Bang?

There are many different aspects of the big bang, but in truth, the glitz of flashy headlines caused the "big bang." Without the story-power of sensationalized headlines, it is unlikely that the term "big bang" would have gotten much attention at all. The News embraced the controversy behind the big bang because the "News" has made its job to be promoting controversy; and regardless of whether or not what is being reported is true or false, controversy stirs up interest and interest sells papers—it's sad, but true.

With regard to the scientific cause of the big bang, the story is either the same or different depending upon your chosen view. For those who do not subscribe to the big bang theory, the big bang was caused by sensationalizing the news to sell papers. While seeming cynical, it is true to the extent spoken of above.

As for what caused the actual big bang (if you believe in the big bang), that is a very open ended question. Science is a study of *how*, not *why*. When we speak about observation, we often say, "**Why** did that occur?" **How** is truly a better perspective for science to use. The term **why** has implications of *intent* where **how** does not, and asking **how** can therefore remain void of *intent*.

Did the big bang spontaneously occur? It is believed that the big bang occurred and re-occurred many times and will continue to re-occur many more times. This leaves us with the question of "how did it start the first time it banged?" In the big bang theory it is believed that all of the matter in existence was compressed into an infinitely small single point, and that it burst forth from intense pressure caused by gravity. Others believe that *all* matter may have been about the size of a basketball. Regardless of which, it would have spontaneously burst forth from the tremendous pressure... theoretically.

This raises a few questions. What was the catalyst that caused it to burst forth the first time? Since all matter would likely have been in a still state, especially if it was an *infinitely* small point, it likely had no appreciable movement and no other matter would have been around to disturb it. The disturbance and/or energy to ignite the bang would had to have to come from within, but the *matter* would likely have been in a constant state for an extended period of time so it would be unlikely to have ever changed. This is unless we make the gross assumption that it had a critical mass point and could no longer be contained, so it suddenly bounced back in the form of an explosion. So much for "infinite".

This is an understandable viewpoint if the matter was dispersed in a fairly even manner and eventually drew together and created an intense gravitational pull until everything collapsed. Anything other than that far exceeds the alleged blind-faith needed to accept the concept of the outside intervention from a Biblical Creator. For all matter to have sprung forth from a single location takes a leap of blind-faith that far outstretches anything that any church has ever asked of its people.

The big bang is somewhat of a copout from doing the work required in finding the truth. And for whatever the reason, the big bang is an effort to hide the truth of the origins of the Universe. That statement might seem to be a strong one, but it is backed up by the animosity that the promoters of the big bang typically have towards *any* alternate theories.

The big bang explains nothing and sets up walls of time against infinity that are too numerous to discuss intelligibly. In other words, whenever someone has questions about the big bang that puts that fundamental belief up for question, then the *time* card is played. The ruler of time is then bent and contorted, and in this case, it is stretched to suit the needs of the preacher of the flawed theory.

When discussing the origins of the Universe, the discussion quickly moves to quantum, then metaphysics, and often ends with the "why?" question, which infers a *purpose* and possibly a *who*. In the end, this is truly the point of contention between people with the issues that surround the debate of the big bang—first *why*, and then *who*. However, in science this sort of questioning is typically not allowed. Yet, if there is not a Creator then there should be no fear with regard to scientifically exploring the Creator avenue because proof will eventually become evident proving either a Creator-based origin or a Creator-less origin.

Die-hard creationists are often equally as biased as die-hard evolutionist big-bangers are. Both groups will take any opportunity to prove their point, often ending in dismal failure for them on either side of the debate when up against real opposition.

Because of the level of order found within each realm, the answer as to whether or not there is a Creator seems evident when we look at all of the "laws" of physics, and when we look at the Universe, and when we look at all of the fossils that we find from all forms of life including plants.

When we find great things we should share those findings with the world. Finding a unique fossil and sharing that with the world is wonderful, and it adds to everyone's understanding. Though, with the big bang it is somewhat different. We have not actually found a big bang. We have observed what appears to be many galaxies, and we see what appear to be massive explosions of stars; but, to date, there is no explicit evidence of a big bang. Claiming red-shift to be the proof of big bang is ignorant of far too many contradicting points of information for it to be taken seriously. And the residual energy believed to be seen in the NASA Wilkinson Microwave Anisotropy Probe (WMAP) is suspect of bias in detecting a temperature differential of only a very small fraction of a degree. If the detected information is accurate then who is to say that it is residual energy is from a big bang? *Finding energy* is different than *knowing what that energy is from.*

What we do have for evidence of our origins are amazing photos of the cosmos that show us various activities that are occurring daily as we see them today from our perspective here on Earth. Everything else that we think we understand about these observations is entirely speculative in the same way that it is with the fantasy based stories that are often invented regarding many of the fossils that have been found.

The pack that follows Darwin's origin theories create entire historical accounts of a creature after finding only a few skeletal fragments, and often only a single fragment. Fantasy stories are made up about the migration, eating habits, and the time period of the creature that supposedly originally inhabited the bone fragment. While it may be possible that the stories are somewhat accurate, it is unlikely because the speculation far outreaches the evidence that was found in such cases. Add to that the fact that these *speculations* are based upon previous *speculations* that in themselves are highly suspect of error. The truth is that someone found a skeleton or a tiny fragment of one, but all the rest is utter fantasy-based speculation that is based on cult-like beliefs. It is claimed to be *science*, but it is not!

Relativity

What is relativity? According to Einstein, the theory of relativity is connected with the fact that motion from the point of view of possible experience always appears as the relative motion of one object with respect to another. In other words, we base everything on the particular observer being discussed— relative to that observer.

Relativity is like re-zeroing the x, y, z coordinates to the point of the observer and then calculating everything from that point. Mathematically this model works very well, but it is short-sighted in many ways.

It is believed that an observer will theoretically see something as shorter when the object is traveling past them, than they would if that item were standing still relative to them. These mathematical games are likely considerably different than reality. Some aspects of relativity can only be in our heads because those aspects are not practical or realistic. Further, if it is, in fact, shorter while we view it in motion, that does not the mean the item actually changes size, light's latency and non-constancy would potentially cause the appearance of size change if space shrinking theory could actually ever be *proven.*

The true relativity in the Universe is our ability to relate, meaning that: What we *think* we see is relative to our experiences and interactions with all of the matter that surrounds us. We create indexes of measure in order to get a personal index on *who, what, where, why, when,* and *how,* and then we use that information to share our findings with others—it really is that simple. While many of our scientific findings are wonderful and useful to all of us, in reality, we do not need them to survive.

We build our indexes as a form of sharing and communicating our findings, and we re-calibrate our counting to accommodate for distances, speeds, and durations that exceed the practicality of our current index of measure—such as measuring a room in meters and open land in kilometers. Let us not overextend the practicality of such indices.

Big Words

The sciences are often filled with unusual words of great complexity, and these words are often thrown around in effort to show our vast individual knowledge so that we can show our scientific "stuff" and how smart we are. Sometimes we have no other adequate means to convey a scientific idea, which leaves us no choice because the big word was likely created to convey the specific idea, but that usage purpose is somewhat rare in pop-science.

In the same way that those who believe they are the "educated elite" utilize their big words, so too will some science elite use big words. Big words might make people sound smart to some folks, but big words actually do very little in advancing science and humanity, or in impressing people who actually understand the material being presented.

It is people's **good ideas** and observations that advance science, and big words are not necessarily needed for good ideas. The simplest form of conveying an idea is the best form, and this includes the words of our language. Take the simple mathematic language form of $e=mc^2$; wrong or right, this simple mathematical word conveys a great deal of information; it is short and to the point.

A good idea is a good idea whether or not there are overly complicated large mathematic or vocabulary words to describe the idea. In the end, to arrive at 4, if I can say in brief $2+2=4$ instead of $2+2-2-2+2+2=2^2$ then I will (please notice the absence of variables.) Pretending that we know it all by using big words gains us little except for an arrogant blind-spot. Science is a wonderful field of discovery, and *sharing* those discoveries and being able to say, "Hey, look at the amazing thing I found everyone!" is the point behind it all. The point of contention is not showing what is found; the point of contention is showing what is *not* found and what is *not* true, as if it *is* found and as if it *is* true.

There are No Accidents

There is a major issue with the big bang that never seems to be addressed anywhere. It is not big bang versus a Creator Creation, but rather this is about the fact that the big bang does not work on *any* level. The big bang has enough flaws in it that any person who is willing to take a moment to study its anomalies will quickly find that the Bible's Genesis chapter one's six days is far more likely to have occurred than the big bang was. The book series, *The Science Of God* explores this in-depth and defies much of what many people perceive western science culture believes about Biblical Creation.

The random spontaneous ignition of the big bang can only be considered a random fluke. To imagine an infinite amount of gravity in an infinitely small point is beyond scientific reason and all reasonable logic and is insisting that all the properties of matter remain while producing intense gravitation. Is it logical to believe that there is any "gravity" left at what is essentially zero point size?

The Universe is far more likely to have coalesced from a somewhat even dispersion of an infinite amount of pre-matter, than to have big-banged and expanded to inconceivable distances in *any* amount of time. The big bang defies the laws of physics in many ways; unless, of course, we ignore the things that don't fit with our demands, and then state that the laws of physics break down. It seems that big bang = big government bucks and much notoriety.

The *order* that we all have witnessed up to this point in the Universe is very consistent, and it does not agree with accidental happening. I call it Modular Inconsistency: The gatherings of matter share the very *consistent* nature of being gathered into clusters in a seemingly *inconsistent* manner. Within those inconsistencies we can see more modular inconsistencies. What this means is that within the appearances of random, there are repeatable general patterns of an inconsistent nature. Spiral galaxies are all spiral, but they are all different and unique in their own way, much like a finger print. All gathered matter is gathered, but each group is gathered in its own unique way.

Nothing is accidental, but predicting high detail with any accuracy is far too complex for us to be able to see it through to the end result due to the incalculable magnitude of the task. Order guides the seen Universe, and it does it consistently, reliably, and very well.

Chapter 18

Our Understanding

Our understanding of our observations is what is being brought to attention in this text. There is nothing more important for any of us to get right than our own understanding of our surroundings, both physical and societal.

A man drove by a house daily and made an assumption that a family with a particular last name lived in that house. The person thought that they had seen the mailbox with a particular name on it each day while driving by it. One day while trying to describe the place, the person mentioned the name, at which point the person was informed that it was another very similar name. And sure enough, the next time driving by the mailbox the realization was made that he had wrongly interpreted what previously had been seen, confusing an "N" with "V."

There is only so much that we can do to be accurate. When someone challenges us, we should not suddenly change our mind, but we should, based upon what they have told us, be eager to investigate our possible error; much like re-evaluating the mailbox. After that we can make the appropriate changes to our thinking—*that* is the true scientific method!

Order

If we are going to idealize anything in science it should be the *order* that we see so eloquently displayed all around us. Light abides by this order, gravity abides by this order, and all matter appears to abide by this order. Even our thoughts want to make order of what we see, but due to our self-serving nature, this is often a challenge for many of us. When we make our observations we are trying to make sense of it all, we are trying to see the order that is inherent in all that we experience. This is proven by the fact that we try to explain things with our math.

Everything that we see is order in all its glory. Our very human quest is for order. If it were not, then we would not have dictionaries and math books. The indexes and standards that we set up to describe the Universe fall short of the order that we see, yet our efforts *are order* nonetheless. The equipment that we use and the experiments that we do are highly ordered, but that still fails in comparison with the order seen in the Universe.

While there is little that we can do to alter the order by which the elements abide, it is an entirely different story when it comes to our own thoughts.

The Capacity to Think

In an effort to understand our surroundings, our thoughts are constantly trying to make sense of what we study. As stated in the previous section, our thoughts are trying to order themselves; but that's where we truly run into problems in science, well... any place for that matter. We cannot defy the order of the cosmos, but we can defy the order of our minds. At some point, our own *freewill* choices are going to be the demise of some of us.

Our ability to think and discover is directly proportional to our capacity to be able to allow order into our thinking. This might seem like an odd thing to say since this is more of a psychology issue than a science issue; but rest assured, there is nothing more important in the advancement of science than the ability to allow order to properly enter into our thinking.

The only true disorder in the cosmos is in our thinking. The acceptance of disorder is acceptance of something that is not true. The only true disorder is something that does not exist. By Webster's definition, *exist* is "*tangible,*" it is touchable. When we invent wild mental explanations that we choose to assign to things that we do not yet fully understand, then we have chosen disorder in our minds—we have chosen to be susceptible to lies. This does not mean that we should not consider a hypothesis, but when a hypothesis has as many flaws in it as a theory such as the big bang or long-age human evolution do, then it's time to abandon the errors, and, at a minimum, re-evaluate the hypotheses; otherwise, we can plan to be crushed by the gravity of the corrective nature of order. True *dis*-order can only be in our minds, and it is our freewill to either choose the path of disorder and to be crushed by it, or to seek to be enlightened by finding what is true.

Scientific Enlightenment

Often when we imagine the thought of someone having a brilliant idea, we picture their idea being a light bulb turning on over their head. When someone obtains the ability to see what is true, then they are enlightened on that particular subject. Scientific enlightenment is no different, but it cannot be achieved without truth—period!

What *is*—*is* what *is*, and that is truth no matter what we choose to imagine about it. It is when we find truth that we can become enlightened. There's not one particular thing that we can do to suddenly become enlightened about *everything*, but we *can* search for the truth about *anything*. However, searching for the truth is no guarantee that we will find the truth. Finding the truth has direct correlation with our willingness to be able to accept things that we do not want to hear when they contradict our current beliefs and view of things. This is especially true when we are ignoring much of the evidence that sits before our very own eyes.

The scientific enlightenment that was predominant in the eighteenth century believed itself to be *the* way, or rather those who followed it believed it was *the* way. Oddly, over time, that, too, became perverted, causing many overreaching theories to be proposed. There's no difference between religion and science; the two are an inseparable whole. And in the end, all of humanity is ultimately seeking the truth.

Humanity's problem with finding the truth and being enlightened with that truth, is that we have, all too often, chosen to believe and follow what amounts to wrong information and lies. Then we go about insisting that our information is the "what is" and that other people must accept our information. Later, when someone challenges our wrong idea, then we hold on tightly to it because it's all that we know and we have invested too much of our life and heart into it to forfeit all of our effort to a new concept of the "truth."

How many times have we seen the denial of additional obvious information throughout history? Far too many! When we have order in our minds, then we have truth; it is when *we* achieve that order that we can understand that we are enlightened on that particular topic. If good and sound information that thwarts our belief is presented to us, and we *refuse* to consider it, then we cannot possibly be enlightened. Truth fears nothing and will embrace any informational challenge.

Order or disorder is our own choice. It is our freewill, and only each one of us can choose it for ourself. When we refuse to embrace having an open mind, then we have decided to live in the dark. We often believe doing so to be exclusive to the Church, but it is not; science is the new predominant offender. True enlightenment can only be considered to be upon us when we have chosen truth. It is our conclusion that is either darkness or enlightened. A conclusion without truth will always leave us in the dark, especially in the field of science.

Let's Sensationalize It!

Going back for a moment to the newsworthiness of a story or idea, sensationalizing is a very good way to get attention. In fact, sensationalizing is the reason that the headlines are written as they are—the flashier the better! Being controversial is the best way to get noticed; though, we must ask, in the long run, does it bring us to where we want to be?

The reason that outlandish hypotheses gain the momentum that they do is due to the gravity that surrounds the controversy. If a person boldly makes an unbelievable statement, it will typically stop people in their tracks, and the people say "what the... what is this person saying?" and *that* is the gravity of controversy.

When these hypotheses are stretched beyond their scope of reality, then they will continue to be spewed, provided no one has the courage to stand up and give contrary and more plausible ideas. The arrogance that usually accompanies these sorts of controversial statements often appear so brazen and confident, that it makes those who actually have the right answers, end up questioning their own brilliant ideas and discoveries. And then they remain silent as they watch in disbelief as the other, erred, theory is spread through the hearts and minds of the general populous like a contagious disease.

When left to brew in the cosmic soup of the media and other entertainment, wrong ideas become indoctrinated into the hearts and minds of the masses of the people. It is when the momentum of a wrong idea has been widely embraced that the science bullies can try to crush anyone who would dare to defy their big bang financial kingdom of prestige. This is not new to the world; this sort of sensationalizing has been happening since as far back as we can tell, so says history.

Order in the Courts

Order exceeds all things known in the Universe. Order is undeniable and it is consistent! The reliability of the order that we see throughout the cosmos is unparalleled by any human achievement thus far.

Every effort that any human makes to communicate screams of order. To be able to come to an agreement on a term so that we can understand each other is order. Our words have order (sort of), our numbers have order, our body language has order, the transitions we believe that we see in fossils have order. Order is everywhere and everything is always working to come into order.

We try to call *order* "chaos", yet we have such a strong desire for order that we all use the same atomic clocks in order to be more harmonious. There is one thing that overrules all of our laws of physics, all of our theories, and all of our hypotheses, and that is **order**. No matter what we say or believe, order will have its way, and all we can do is to observe its beauty and share those observations and explanations with the rest of the world! If only we would recognize this order when debating such topics in the courts, we could be living in a very different world today.

The New Constant

In science we must be cautious when ascribing the idea of a *constant* to another idea. Order is one of the rare constants in the Universe, and while we do not fully grasp the magnitude or power of order, it is far more constant than light or any other constant that we have ever established. Science must adopt a new constant in the scientific mindset, and that constant is *order*. Though, there is another constant that seems prominent, and it is our constant nature to doubt; maybe we can make our vast and foolish ability—to doubt what is true—to be our new constant. But I digress.

Creating the Index

All means of index are a form of communication, and communication is order. The only reason that we communicate is to either bring order or disorder to others. We either want to share ourselves and our discoveries with other people, or in our arrogance, we want to destroy them even when they are correct.

When we choose to communicate in an effort to share, we do so with order being the primary directive. We create language, words, and symbols (indexes) so that we can share what we see and discover with those around us. We might be sharing information about our day at work and how we felt about it, or we might be sharing the *information* from, or about, the work itself and how we feel about it.

With our indexes, we seek confirmation from those around us about our findings, but it's not really our findings we want confirmed, it is our analysis about those findings that we want confirmed. When we share our thoughts and our understanding, it's easy to be personally crushed by others who don't agree with our new information and conclusions. New information means a new index, and having to adopt a new index is like going home and having the lights out in your house after someone had secretly rearranged your furniture in a very unfamiliar and seemingly illogical order—but that doesn't mean that it's not a better arrangement. We might trip over our furniture at first until we become accustomed to its new arrangement or until we turn on the lights.

Written documentation of observations is our index, and all that we know is relative to those indexes. Correcting them as soon as and as accurately as possible will allow us to see the ever bigger picture that awaits our discovering.

Chapter 19

Weighing the Facts

The facts are the facts, right? No, not necessarily. It seems that the term "*fact*" is somewhat subject to each person's interpretation. Often we believe a *fact* to be true, and by definition I suppose that it is; however, this is another word that seems to get bent to fit our own individual understanding of what we see, or what we want to see be correct.

I have heard people utter that long-age evolution and big bang are "facts." Some people may find this comment to be short-sighted, but it simply is not "*fact.*" Since we were not there while long-age evolution supposedly occurred, we cannot say that it is fact. It is a fact that we found fossils of diverse size and form, but as to the ages of those fossils, that is utter speculation. Some scientists might say that the carbon-dating and atomic half-life is an accurate way of dating fossils, but here again is a problem. It is a fact that we date fossils with carbon dating or other similar methods, but the accuracy is left wanting, especially when considering the margin of error on items that we actually know the ages of. As far as the calibration is concerned, that is an entirely different matter. We simply cannot know for sure that our calibration is correct for extremely long ages. When we take a small slice of life and use it based upon our only possible modern timeframe in which we have lived and observed, then we are doing a tremendous amount of speculation when we project those dates to millions and even billions of years. This sort of extensive extrapolation is what is done with radiometric dating. Radiometric dating methods have a margin for error so large that they are rendered useless for any short-term dating. Our perception of atoms is influenced by our scientific experiences and the words of others in the science community. If we are wrong about light, then we are likely wrong about the way we count when we date fossils.

On the surface of our understanding, we believe all matter to have behaved the same as we see it today while we live and breathe, but that does not mean the atoms now used in testing our ideas behaved the same way for the last x-million or x-billion years, especially if we are going to claim singularity. Stating that radiometric dating is factual in its accuracy is not quite accurate. Again, it is a *fact* that we radiometrically date things, but it is not a fact that it is accurate to the calibration that we have assigned to it when dating millions and billions of years. Radiometric dating can only be accurate to our *perception* of our chosen index. As mentioned earlier, school children are more accurate with their ninety-nine cent rulers, than science is in many aspects, including radiometric dating. Most people simply do not question what the "experts" say, and will go on to accept radiometric dating methods as highly accurate, even though these dating methods have an excessively wide margin of error. There is no other industry where you get a pass on such vast inaccuracy. Even crime-rings whose business is dishonesty are more accurate when counting than is big bang theology.

Extending radiometric dating beyond **our** scope of time of experience, is similar to judging the weather by a two-week slice of time. Let's say that it has rained for the past two weeks and was overcast and cloudy. Is it then fair for us to take that same experience and to extend it thousands of years back, and then say that it has been cloudy and rainy at that location, constantly for several thousand continuous years? The *facts* are the *facts*, and a better understanding of what part of something is actually a "*fact*" will allow us to take things at "face value" much more accurately.

If something is made and then exactly two weeks later it is radiometrically dated, the date ranges given when dating the item are typically far out of proportion, often to hundreds and even many thousands of years. Yet, we believe radiometric dating to be accurate with much older items when it is likely *not* any more accurate, and may even be exponentially more inaccurate the older something is. If our abundant radiometric dating errors for recently made items are inaccurate, then we can most certainly speculate that the same massive percentage margin of error is present in dating ancient organic matter.

Comparison

To compare something, is to measure it. Comparison is a reference point; it is the way in which all humans communicate. Comparison is index; it is comparison in reference to where you have been. It shows us where we are, and then we can see where we are going.

The indexes we create are indexes that are built upon our best observations of internal comparisons. For instance, how gravity interacts with a feather, in comparison to, how gravity interacts with a rock. Once we record these observations or findings, and build our index, then we share these findings with our friends, our peers, and the world.

Thinking at the Speed of Light

When well-spoken and articulate, quick-minded people deliver their conclusions about their observations, typically their view is well laid out and easy for others to understand. Science often has a tendency to obscure hypothetical ideas with complicated mathematics that only skilled mathematicians can unfurl. Of course this does not make the hypotheses incorrect, but it does remove a great deal of potential contrast to those ideas from other people, which is due to the fact many people are not learned in those types of mathematics. Not to mention that if something is wrong, then as a technicality, it cannot be understood.

In many cases, it is those who have a well-laid plan or theory who have become the face of science. *Thinking at the Speed of Light* is important for anyone who is going to be facing potential opposition of their ideas. Sadly, in our headline-hungry world, accuracy has little to do with anything. Fast thinking, well-studied, articulate, outspoken representatives are going to be seen as correct even when they are utterly wrong, and often their errors do not become apparent until after their time in the spotlight, or even until after their deaths.

Sounding smart seems to be all that you need in order to force through a theory that is highly suspect of error. Validity seems to have little weight in the popularity contest of sensationalized science.

We Built this Universe on Math Games

Our current scientific model of the Universe is largely built on mathematic formulas that step eons beyond their scope of reality. The size and distance of stars, the ages of the celestial bodies, and the ability to be able to compress matter to infinitely small sizes, are all built upon math games that work similarly on the mind like the optical games shown in the following Figure 13 Mind Games, Page 235.

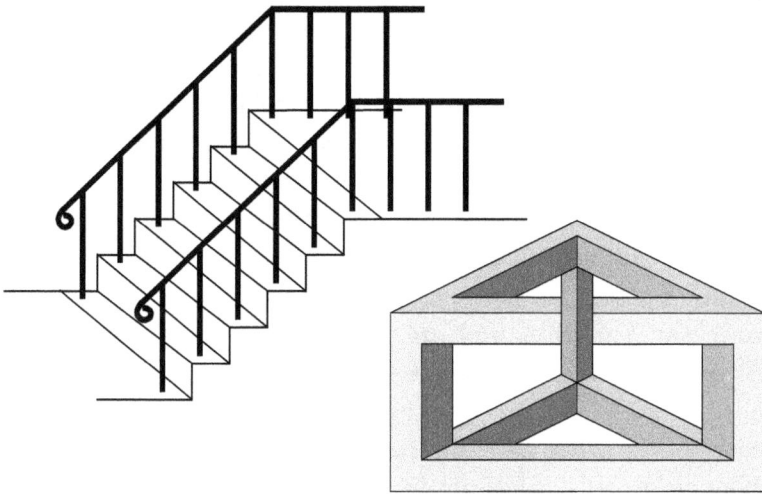

Figure 13 Mind Games

Nowhere are these math games more prominent than in time and distance calculations, or within the compression and size calculations used in quantum physics. When observing math games of space travel, we are often entertained by percentage and squaring tricks used with regard to the observed versus the observer. These are wonderful mind exercises, but depending upon how and why they are being used, the extension beyond the physically measurable scope of these methods is quite deceptive.

Geometry and math do not dictate Creation, they simply measure its order and consistency and help us to try to describe and understand it. Measuring is little more than a lacking comparative index that is used in our attempts to understand what we see. While we can draw illustrations on paper similar to those just shown, in Figure 13 Mind Games, Page 235, we cannot produce them in reality as they appear in our mind and on paper. The same principle holds true with our scientific math.

Building the Index

Time, distance, hour, meter, and truth are all indexes; they are truths. But they do not determine truth, they only measure. If

someone changes a ruler's meter-distance, then it cannot be
trusted and it is a lie unless we can all come upon an agreement
with regard to its new value. If we do redefine what a meter is,
then we will need to re-measure everything because the items
that we measure will not change size with our recalibration of the
length of our meter. The index of truth is something entirely
different, but we often fail to see it that way. For many minds
"truth" is subjective because we believe that we can have your
truth, and my truth, and his truth, and her truth; but is this
correct? No, there is only one truth and that truth is the single
index that is never changing. Truth will always be truth
regardless of what we claim it to be. We can deceive ourselves
about truth, but it will gain us nothing except error. No amount
of trying, screaming, fighting, killing, demanding, sensationalizing,
dazzling, or fast-talk will ever change what is true.

Truth is an index so accurate that it is infinite. All of our
efforts to get our own way in any field of work can never destroy
the actual truth, but truth has power to destroy anything that is
not true. In the end, theories such as the big bang will be proven
to be either true or false, truth or lies; and nothing that we say or
do will ever remove that simple truth.

Take the concept of Heaven, with the angels and pearly
gates and all; nothing that we say or do is going to change what is
true about this. We can deny that it exists, or we can demand
that it exists all we want to, but in the end, what *is*—*is* what *is*
and we will find out when we do or do not get to the pearly gates.
What we think about Heaven, now while we are here and alive,
is simply irrelevant as to whether or not "Heaven" exists.
Although, it is believed by some that denying the truth of heaven
will cost an eternity in hell. In the end, we will all see, or not see,
because it either is or it is not. If Heaven *is*, then many of us are
in for a terrible awakening, if it is not, then it simply does not
matter.

Regardless of our personal choice about "Heaven," or any
scientific belief, in the end, we will be held accountable for our

beliefs. With Heaven, it simply won't matter if Heaven does not exist. However, with science and things like the big bang, if any of these sorts of far reaching theories are disproved, then there will be a considerable bit of embarrassment, humiliation, ridicule, and waste of life's work and focus for people to endure who had the wrong "facts" and led other people astray with those supposed "facts."

What is Time?

The indexes that we adopt are our own choice and we will bear the consequence of our choice whether good or bad. Our first point of focus should always be the measure, or index, of our communication. Misunderstanding the idea behind an idea label such as "*time*" is costly to the misunderstanding-perceiver's productivity. *Time* is an idea that we use to measure duration. Duration is an intangible unspecified existence interval. For instance:

From now...

one...

two...

three...

Until now!

We cannot bend or alter the period of time no matter what we do, it simply can never change. We can, however, change the calibration of a measuring device such as a clock; for instance, by changing the term hour from being an equal division of twenty-four segments of a single rotation of Earth and making it be fifteen equal segments of the Earth's rotation. Changing the index of the duration of an hour does not change the term hour, it changes the value that we assign to the term hour, but the duration of the rotation of the Earth itself stays the same. In science, however, we change the duration of an hour and then we believe we have changed time and space.

Remember:

Gallon is to *volume,* as *hour* is to *time.*

Volume is to *water,* as *time* is to... what? Duration? Not quite

Volume is to *water,* as *time* is to *existence*, or how long something has *been.*

"*Time*" is a measurement of duration of *existence.* We can call time a dimension, but then there are also many more dimensions, such as volume, circumference, temperature, color, shape, texture, etc... these are all properties, as are length, width, and height.

How long something was there or how long something took is a measurement of the of duration of existence, and we measure it in increments of duration of time just like we measure space in increments of distance, and the substance in that space in increments of volume.

The Truth about Science

Science is not an *exact* science. We have led ourselves to believe that if we divide something to ever smaller increments that we are being more accurate. The finer the resolution of an index, then the more specific we can be. Right? No, this is not right at all! If the entire world has a ruler and calls it a meter and everyone's meter is the same, except yours, then you can divide your meter in as fine of increments as you would like to, but it will still be different than the rest. When you try to communicate your measurements that go by the same name of "*meter*" as everyone else's meter, but is calibrated differently, and is actually a different length, it will cause confusion because your *meter* is longer. You will always feel short-changed when doing commerce with the rest of the world because you will be expecting more, and everyone will give you less than you expected. And in your mind, everyone will equally short you which is due to the fact that your meter-stick is different than all the rest.

Science is only as exact as our willingness to share identical index calibrations; and our index calibrations are only as exact as our ability to properly understand what we are observing. There is no escape from this basic logic. If we are misunderstanding what we observe, then higher precision will only be a higher precision of **in**accuracy.

We have been building a premise-on-premise physics model that is ultimately paralyzing and is dangerous to science. Our scientific house of cards *will* fall and *will* need to be rebuilt if its foundation is not repaired soon.

We have allowed society to somehow come under government control, and we are now *required* to teach big bang and long-age evolution to undiscerning children from grade school age through college age. And though it may be referred to as "theory" in words, it is taught as factual in attitude. This government sanctioned assault on science is having a tragic effect on the ability of the upcoming scientists to be open to truth. And it is turning many children away from science. This is identical to the way the church leaders inhibited the followers from seeing the truth during the lifetime of Galileo and others.

Just as the church should remain separate from the government, so too should science remain separate from the government. The new government sanctioned religion of science teaches two fundamental principles that have more flaws than the church's doctrine ever did or could be imagined to have, and likely more than it ever will. Anyone attempting to teach any view opposing long-age evolution or the big bang will quickly be attacked for their opposition. Funding can get cut and police can be called, and people can be jailed for proposing any view that does not line up with this new religion, this is especially obvious in the public school systems.

The temples of science are now more revered than the churches were, and they are paid for by the government using the tax dollars of the people. Museums of natural "history" do not just show fossils and say that "we speculate"—no, not at all. And the public schools do not teach that Darwinian-evolution and big bang are unsubstantiated theories. The contemporary views on "natural history" are taught as "fact", as if the time frames and Darwinian model are what is; which is all dominantly paid for by the government's spending of taxpayer funds.

There is no hard evidence of the vast time-frames perpetrated on the unsuspecting public; and, often, when the teachers do not force the children to pay homage to the new religion and worship the patriarchs who are the likes of Darwin, Hubble, Sagan, and Hawkins, then there is a heavy price to pay for that teacher and possibly a heavy price for their entire school. A student will typically be shut down by a teacher if their view does not agree with the textbooks. I am not referring to religious beliefs being offered, but rather *anything* that conflicts with the government sanctioned science agenda. In the public schools no acceptable alternate is allowed to be discussed. It is either, long-age evolution and the big bang, or it will be ridicule and threats of loss of funding, penalties, lawsuits, and yes, even imprisonment! If that is science, then *true* scientists should want no part of it.

The truth about science is that any event or phenomenon that is not specifically observable is not "proven", and it cannot be proven until we find observable evidence with *no* errors. Until then, we are only guessing in our loose estimates of our origins, *including* guessing about time and space. When we can see the actual light waves, then we will know for sure that it is waves that we believe to be apparent in the double-slit experiment. Until then, the jury is still out on light-waves along with so many other "facts" of science. The science ruler has been tightly bent to a kink.

The Balance of Order

Our science errors, whether large or small, will eventually come to light when the truth is found. Eventually, the truth **will** be found! Our arrogance and effort to suppress true new information that violates our old scientific laws has had a considerable negative effect on the minds of the up and coming scientists. Eventually one young scientist will start a new revolution. He or she will reveal to the world our errors that we have made over the last several hundred years. Certainly these errors have brought us forward in some aspects of understanding, but in other areas we have been greatly inhibited and blinded by those errors.

There is an equilibrium of order that cannot be defeated in the long run, and nothing that we do will ever stop truth. Eventually, all of our errors will be made straight when some new fresh, young, bold mind comes along and sees the holes in the outdated, error-prone science "laws" of the twentieth century. Then this will bring upon the people, a light unlike anything we have yet to discover.

The balance of equilibrium is order, and order will always prevail. You do not have to like order, and you do not have to agree with order; that is each our own choice. But I assure you this one simple truth: The truth will prevail and anyone who speaks against the truth and denies the truth will pay a very dear price. Truth *will* have its way with us in the end.

We can shout from the tree tops that two and two is five, but no matter what we do, or what value we assign to the symbol "4", two plus two will always be this many items→ I I I I

The Lies in Peer Review

Societally, we have created a scientific monster that may be difficult to unravel. Our method of scientific acceptance is peer review. If a hypothesis is presented and reviewed and accepted by peers, then we believe that it is worthy to be published in various science journals. When it has been published in the journals, then it is reviewed by more peers, and so we believe that it is sanctioned by science and is therefore true. We might not think of it in those terms, but this is essentially what occurs.

Peer review has one glaring problem: The peers all believe the same thing and have come to power. Just as politics is swayed and will lean left or right when one party dominates all branches of government, so, too, is science swayed. This problem is worse in science because no one can get voted out. Must we endure the monarchy as their incorrect proclamations are set in the stone-cold hearts of their minions until the kings of science retire or die? Or will we take back science and seek what is ***true***?

Peer review is useless if we only get approval from those who agree with us. ***True*** peer review is when, based upon the evidence you have found, you can convince, without coercion, those who normally do not agree with your conclusions. When someone says, "Hey, there's a hole in your parachute!" we had better listen to them, or be prepared to plummet to our scientific doom. Having our peers tell us that there is *not* a hole in our parachute will not save our life when the mass of our body meets with the mass of the Earth from the gravitational pull of truth. Order will have its way with us and we will plummet into the abyss if we insist on defying truth.

Chapter 20

Which Road Should Science Take?

Whether or not we choose to believe it, most everything in life is a choice. If we are working with a difficult person that is bent on insisting that their wrong information is correct, then we have the option to walk away and state our hopefully accurate case elsewhere. All too often we tolerate the bad behavior of others, and doing so allows that sort of bad behavior to flourish. I once heard a very wise saying that goes something to the order of "What you are willing to tolerate will not depart from you." The scourge of mankind, in all areas of life, is our own *fear* of conflict. We mistakenly confuse the screaming, irate person who demands their way, with a fed-up person who demands the truth.

Life is a choice, and we each must decide for ourselves whether we want to choose to be wrong, or be true. Collectively, many people and leaders in the church chose an unproductive path in past centuries, and we wrongly believed that society had learned from the church's error. The critics of the church eventually spoke up, but then never shut up. Humanity has an inherent problem in the area of right and wrong, or correct and incorrect. Often, a group of daring people will see errors and then stand firm and speak out against those errors. The silent others, who see the same errors, will then rally to the side of the rebels once they believe that all is safe and clear. Then when the rebels come to power they impress their beliefs on the people with the same or greater vigor than their defeated counterparts did. Then eventually they add more errors to their defeated counterparts' beliefs.

History has been repeated many times with a coup of rebellion and the subsequent replacement of a new set of adjusted, but equally inaccurate, laws. Nearly every turn of power has shared this same pattern. It does not matter if it is in the school yard, in the work place, in the Church, in the government, or in science, because the story is always the same.

Which road should science take? The road to open-minded research of what is true, of course!

Order or Chaos?

Everything in life is a choice; those choices might not meet with our approval, but they are choices nonetheless. If you want blue and the only choices are red and green, then you get to choose red, green, or none. This sort of thinking does not resonate well with many people. Regardless of which realm of life we look at, we humans, in general, want things our own way.

The populous, as a whole, has somewhat disconnected science from humanity. A perspective mostly unrealized by us all, including scientists, is that we tend to believe scientists to be robotic in nature. Meaning that somehow we view scientists as unbiased machines that will only seek to figure out how it all *truly* works. But, this is not the case at all; scientists are humans and have all of the identical flaws that the rest of humanity has. This means that unjust agendas *will* be driven through for prestige, fame, money, power, to save face, and for plain old stubbornness and ignorance.

Science is a choice, as is all of life. It is a choice of order or chaos. When we choose order, then we have chosen to seek out only the truth. Truth appears to us to be a fickle thing because we don't know truth very well. Truth is what we find when we reveal it by removing all of that which hides it. Most of what hides truth is in our minds. Earlier when we discussed order and disorder, it was pointed out that the only real disorder is in our minds—disorder **can only** be in our minds.

The Universe is highly ordered, and, as far as we see, it is repeatable into the multi-trillions and beyond on a massive scale; and it is repeatable to unfathomable quantities on the micro scale. The repeatability of order is the single strongest visible phenomenon that we are aware of. We either choose to see order and describe it, *or* we choose to rationalize inferior explanations because we are frustrated with our own inability to see how order is affecting our particular area of study.

Choosing a Scientific Path

There is only one path for true science, and any other path chosen is not true science. We have taken the wrong path many times over the centuries, and, with regard to science, our erred path was due to our poor interpretation of our observations. We did it again in the nineteenth and twentieth centuries, and now we are still reeling from doing so over a hundred years later.

We get to choose the path, to shine or shun; truth or incorrect information—it is each our own choice. Science will either shine with the truth, or shun the truth for something that makes us feel better for the moment. The only real sin in the world is to deny the truth of what we see; and in science there is no sin more heinous than doing so by claiming that we are using the "scientific method" when we clearly are not.

The scoffing that we witness when someone proposes any other idea than the big bang and long-age evolution is so common that we barely even notice it any more. I find that the best information is almost always derived from the opposition's camp. When we cannot see our own hypothesis clear to truth, then taking the time to consider what those who oppose our ideas think, will often help us to sort things out and find the truth.

Science's sole true purpose is to try to make sense of what we see, and then to be able to explain it so that we can reproduce reliable and predictable results; but merely repeating reliable and predictable results does not make our hypothesis correct. The undefeatable integrity of *order* allows our wrong ideas to appear right. And as long as our narrow-viewed hypothesis has a consistent nature, then it will coincide parallel to the order of the way things *actually* work **within the hypotheses' scope**. We could consider this like driving in the ditch along the roadway. Just because you can get to the same place does not mean that you are doing it right. Just because the end result is correct does not mean that our formula is correct (again, consider addition to get 2+2=4 by incorrectly using 2x2=4); also we may experience an unneeded bumpy ride along the way while riding in the ditch. Some people might enjoy the ride, but it is not the best route, and it will damage the property of others on the way. We can also drive at a speed different from the speed limit and still get to our destination, but the time traveling will be different than if we follow the speed limit taking the same route.

Our formulas are perhaps our most shortsighted pieces of work because they are **all** reverse engineered. In science, we do things backwards; we look at a result and say, how did this happen. In a creative mode that is not possible; if a person wants to create something that has never been, then they must picture it in their mind, and then go about figuring out how they will make the thing come to be. They cannot dismantle it and study it and then reverse engineer, as it occurs in science.

Quantum Mechanics

No matter what we think, we cannot ever be certain about anything unless we have chosen to seek the truth about order. And even at that, we must be always and ever-willing to correct any errors in our thinking. You cannot be certain if you do not choose truth, not *the* truth about a specific thing, but rather *Truth* itself. Quantum mechanics claims randomness, but then claims accuracy; this gets back to that issue of measurement and dividing something that is wrong into even smaller increments. Wrong is wrong no matter how finely we increment it.

Believing a photon to be in two places at one time does not compute, and is not logical and it is likely wrong or greatly misunderstood. Since the double-slit experiment appears to have shown such a phenomenon, and it defies the logic of our real-life experience, it seems that much more experimentation is needed before we can draw any strong conclusions. In other words, we are missing something, so instead we ascribe what we see as what comes down to magic. Here is the core of the problem that science faces: conclusions are being drawn with little substantiation, and then are promoted to unsuspecting and trusting people as if those conclusions are "fact".

In science, we must make up our minds on the hypotheses for quantum mechanics; either we are going to believe in things that are counter intuitive, or we are not. Certainly, all avenues should be explored and analyzed, but the final conclusions to be promoted as "fact" should wait until there are no obviously apparent errors in the hypothesis.

Deciding what is True

It's apparent that most people, including scientists, do not realize that each one of us gets to decide in our own mind as to what we *think* is true. There's an assumption that we *must* believe what other people tell us; yet, every one of us is accountable for our own actions *and* thoughts. In the end, we cannot blame others for our errors. Though, we do each hold some responsibility for leading others down the wrong path. But still, in the end, it is each our own choice of what we will choose to believe.

Whether the children in the school system are young or old, they have been taught to trust their instructors. Parents send their unsuspecting children away to school expecting that they will be shown only good and true things. Instead, the hearts and minds of the children are often filled with agenda driven information that typically contains political undertones. Then any brave children who attempt to question that thinking are shut down with ridicule and told that they are wrong by their superiors, as well as by any gullible fellow students.

Early in the twenty first century, things came to a point where any student who wore a shirt proclaiming big bang would be credited for wearing the shirt, but another student who would wear a shirt equally proclaiming Biblical Creation would be threatened with suspension and punishment.

Just because the church says that there is a Creator, or just because many scientists say that there is not a Creator, does not make either case so. Just because science says there was a big bang, or just because some churches say there was not a big bang, does not make either so. When all is said and done, whoever is correct is correct, and no matter what we believe, that true fact will not change—ever!

Our correctness *is* dependent upon what *is*, but what *is* does *not* depend upon our correctness *or* what we think. Though we act as if this is so, the truth is that science does not get to choose what is true. Instead, science only studies what is true and makes a best, but often poor, attempt to explain that truth.

Truth is a personal choice. We do not get to decide if something is true. We get to decide if we will agree with and admit to what is true while we are compiling our theories as we live and breathe.

Chapter 21

Beyond the Infinite

The speed of light, time-travel, gravity, time, and the big bang are prominent areas with a great ability to deceive us. For many people this message might not be popular, but know this: It is those who refuse to consider this message, about scientific truth, whose ways we must reject.

The big bang has outlived its usefulness to anyone with an open mind, but this does not mean that we must suddenly embrace another wrong doctrine. Rather, we should seek to understand how it all came to be from other perspectives. Maybe some of those perspectives include a Creator, and maybe some of them do not include a Creator, but one thing holds true—utter rejection of anything that cannot be disproved is foolish and dangerous to the rejecting individual.

If someone can present convincing evidence of a big bang then that would be good, but they will need to be able to explain *all* of the vast inconsistencies in their belief. Let us use the same standard for science that science held the church to over the centuries. Darwin's stretching of age in order to prove his theories led to an international effort to prove long-age, which in turn bolstered the big bang expanding Universe theories.

It would be foolish for anyone to state definitively that the Universe is not expanding, but then the word "*expand*" needs to be clearly defined. Things can expand in more than one way. There is explosive expansion as postulated in the big bang, and there is also inflation or stretch expansion, and then there is growth expansion like in a *growing tree*.

Maybe the Universe is in a constant process of Creation and is growing in size infinitely and forever, and new matter is being Created as we speak and is coalescing into new celestial bodies. Our narrow minded, paralyzing big bang assumptions are short-sighted and show that we have a difficult time imagining *infinite*.

Know this: When we think that we have reached the end of the Universe, or that we can see the edge of the Universe, or when we think that there are multiple universes, or cyclical big bangs, then we have set up walls to infinity. This is our short-sighted human way of dealing with our inability to imagine the infinite: We want to place limits on infinity, and then later we realize there must be more so we expand our thinking, only to build yet another mental wall for infinity to have to once again break through.

Today we believe that long past church leaders had limited Creation to six 24-hour days, but now it is *our* science that is limiting Creation. What was once six thousand years is now limited to 13.7 billion years, respectively. You run the risk of losing credibility in the scientific community and you risk losing your job if you defy this consensus. Science does not run the risk of becoming like the church of the past, because science **has** become the church of the past!

The True Age of the Universe

The twentieth century put the birth date of the Universe at 13.7 billion years ago. Is this correct? No one truly knows because it is all utter speculation that is based upon some frighteningly variable indices. If our day-to-day lives functioned as irregularly as the mathematics and theory system of science, then most scientists would not make it to work in the morning. Luckily for science, industry has a better fix on practical application, and uses *reliable* standards that allow for our vehicles to stop and start dependably. This way everyone can get to work safely, unless of course someone *bends the rules* and decides to fail to use their brakes and subsequently goes through a red light when *you* are in the intersection.

We may never know the true age of the Universe, but we can certainly do better than to imagine that it is about as old as we believe the most distance galaxy to be from us in light years. Again, it needs to be pointed out that a galaxy that has its light hitting our telescopes today, had to be in that approximate location when the light was emitted at the hypothetical 13.2 billion years ago. This is based upon our current estimation of the speed of light. This means that the observed galaxy, if moving away from a big bang ignition point, would be considerably further away today than it appears to us. I challenge anyone who questions these things to re-read the sections that discuss the age of the Universe and think upon those points. See if you can honestly say without question that the big bang had to have occurred, after you have considered *all* of the inconsistencies that the big bang theory contains, which are not all listed in this book.

From a different perspective, if the Universe is in a state of collapse, then what is the progression of that collapse today? Is it much further than the light we see from the alleged expansion, or is it much nearer to us? If it is nearer, is that possible? What happens to light if the light's source is moving towards us for 13.2 billion years, and at a rate that is possibly faster than the speed of light? Will we be surprised one day from a sneak attack from a collapsing Universe? Too many possibilities to consider I suppose, so we disregard them because none of them make much logical or practical sense.

The perversion of the term "science" has become so prominent that anyone attempting to thwart the big bang is discredited as a fundamentalist Christian who believes that the Earth is flat even if that person is not a believer in the Bible or a God or a god. The big bang and Darwin's evolution have led us to wonderful discoveries only because we actually began to look for new things; but they have also potentially brought us to some very flawed conclusions that will likely keep us stalled from progressing in finding out how it all *really* came to be and it will also keep us from serious space travel.

We should always keep looking for answers. However, in truth, it is unlikely that we will ever know the age of the Universe with any certainty. This is unless we are someday able to travel at many billions times the speed of light. But, this is unlikely to ever occur with the current mentality because we have chosen to believe that Einstein "proved" that exceeding light's speed *cannot* be done at all.

We have a fantasy belief that time travel is possible, but we refuse to believe that exceeding the speed of light is possible. To all of you young up-and-coming scientists and inventors, please understand that there is a Universe full of wonder and order just waiting for *you* to cast off the old foolish limits of the early twenty-first century, and then discover and understand the vast treasures in the Universe. The true rewards will far outshine the science of the nineteenth, twentieth, and early twenty-first centuries. The moment we believe that we know it all, is the moment we can no longer learn. When you understand this point, then you can learn *anything*. Follow vain pursuits if you will and insist on time travel, bending space-time, and a big bang; *or* see through the errors and dream the possible!

"*Science is in the Doubt Business*," but is It?

I once heard a scientist say that "science is in the doubt business," but that is contrary to the meaning of the term science. The word **science** means "*to know.*" If we know something then we do not need to doubt it, though we can check our findings when contrary thought arises. But, to be in a constant state of doubt is not the way to conduct science.

The truth of the matter is that *science* is in the truth business, and science should not believe every nonsensical whim that comes along, but science should at least process the information honestly. The whole point of science is to discover the truth. Did you get that "*dis*"- "*cover?*" That's right, in science we try to *uncover* the truth from the blindness of our own minds. The answers sit before our very own eyes and they always have. It's just that *we* fail to see the answers because we have chosen to believe wrong information.

Science is not in the doubt business; science is in the *understanding* business. If we believe that we cannot exceed the speed of light, then we *will not* exceed the speed of light because we won't even try. It is said that there was a time where some people felt that we would burst into flames if we exceeded speeds that we now travel every day with our vehicles. There was also a time when some people thought that we could not exceed the speed of sound, but these myths have been proven wrong. In the same way that we can exceed the speed of sound and its relativity, so too is it likely that we can exceed the speed of light and its apparent relativity.

The distances are so vast that if the Universe is at all in motion and if it is moving at the speed of light we would barely be able to detect this. From a perspective of the observer's relative speed, we cannot exceed the speed of light because *we* always recalibrate our x, y, z to the source observer, which in this case is us. In reality, it may be dependent upon the relative movement between the observer's motion and the motion of the substance through which the light is traveling, a substance which may or may not be there, and which may or may not be moving.

Imagine the possibilities of being able to travel at speeds anywhere near the speed of light...

We send skyscraper-sized rockets into space with controlled burn bombs because we believe that is the only way to break free of Earth's gravity, but is it? There *is* a better way waiting for us to find it. The likes of Abraham, Copernicus, Galileo, Newton, Kepler, Da Vinci, Wright, Tesla and many more did not sit around believing the drivel that was presented to them by their contemporaries. They all had one thing in common: They saw the world for what it is and they did their best to cut through the myths spewed to them by their contemporary counterparts about their passion.

Their bold research in opposition to "common wisdom" was not a result of foolish stubborn men. Rather, it was the result of men who confronted the truth to expose it to the world by stripping away the myths and lies that they saw in the conventional wisdom.

You may find it odd that Abraham is mentioned with the others who are known more for their *scientific* contributions. The choice to believe the Biblical Abraham account is an individual and personal choice, but in the historical account of Abraham, he defied his father who sold carved idols that people worshipped and sought their sustenance from. His defiance was a result of his ability to scientifically look through the nonsense that we are led to believe. Regardless of your take on the validity of this account, the lesson is the same. Abraham questioned and said "could that which created everything be a statue?" Then he went on to question about the Moon and the Sun as the source of all things, and in the end concluded that a Creator must be greater than the Creation. In the account of Abraham, to draw his conclusions, Abraham used the logic that most of us are born with, which is essentially this: If there is a Creator, then that

Creator must be far more dependable, and far more, than what we are able to see.

Most of the true science patriarchs made their discoveries in the face of the church and scientific consensus, and they did it due to their belief in the consistency of their Creator. They were men with deep conviction whose minds were not tightly bound to conventional wisdom. They also did not adopt false beliefs about that which they saw. Their earnest efforts were to discover for themselves, and then to be able to explain what they observed to others. They did not seek to dictate what their observations would do to the environment as we do today with our interpretation of "*law*".

Infinity

One thing is certain: As a whole, humanity's ability to think is limited only by our beliefs and by our ability to perpetuate our kind. When we lose that, then our infinity has ended. Maybe infinity is really a choice of something to believe, meaning: If you are unable to imagine it, then for you infinity cannot exist, but infinity still exists for those who can grasp it. This is not some metaphorical statement, but rather an observational truth about us humans. If we do not believe that something can be done, then we typically will never try. Most people don't ponder infinity, and if they ever do, then they likely only do so for a brief moment and quickly move on because it is more than they want to mentally process.

If you were dropped into the middle of an ocean and did not believe that there was a hidden underwater mountaintop a few inches below your feet, then you would not know enough to set your toes down and you would likely never try, and eventually you would meet your end in the abyss. Though, if someone told you that the underwater mountain was there, then you would likely not stop trying until you found it so that you could stay alive.

Bending The Ruler to suit our minds will get us nowhere. It's great that we have sent rockets into space, which was done based upon our prior knowledge; and we have learned a great deal from it. But how much better would it be to find a less expensive, safer, more efficient, and a faster means of doing so?

What we call science is a wonderful field for up-and-coming science-minded youth to pursue. It is my intention to get those who have the true gift of *proper* understanding to pursue science and reveal the wonders of the cosmos to all of humanity. By realizing the blind arrogances that we have been discussed here, and then overcoming them, we can accomplish this goal.

The most important message that I want young science-minded people to understand is that science is **far** from having it all figured out. There are enough discoveries left to last infinite lifetimes. Just as we have discovered many of the places on the earth, so too we may be able to do the same in the *Expanse* of space. But please understand that there is still much left to explore even here on our planet Earth. Let us never cease to understand our own home.

While it is good to have large, expensive, and high precision equipment, it must be acknowledged that most of humanity's important advancing discoveries that are used to this day were not done with huge telescopes or in multi-billion-dollar collider-labs. Most great advancements were done, and still are, by people who immerse themselves in finding the truth using little more than pen, paper, and mind—even if finding the truth means that their previous thinking is proven wrong. Also, these people spent much of their time in deep thought wondering how it all came to be. They didn't want to make up fantastical stories of big bangs and evolving amoeba and entire fantasy stories about the fantasy creature based upon a single bone fragment or a part of a tooth. They just wanted the truth *regardless* of what that truth meant to their own beliefs.

It might seem simple to understand that people just want the truth, but that is not the human way. Our self-centered nature causes us to have the inclination to want it *our own way*, rather than what is *true* and *best*.

If science were a human being, then we could say that the field of scientific discovery is not yet even born and that it is still in its mother's womb. We certainly should not abort it because it is still developing, and with proper care and nurturing it will birth a wonderful field and will lead us to a promising future that will offer humanity unlimited possibilities.

The generation that grasps the true nature of order gets to give birth to the *true* field of science, and when this occurs it will not be because of accidental discoveries.

The mysteries about to be revealed by *True* science are vast and wonderful, as is vaguely evident by what we see in the Heavens. Be a part of this discovery and place your name forever in the minds of the human race. Do not do it because you want glory; rather, do it because you want to know what is truly occurring. When you discover the wonders of order, then share your discoveries with the world so that all of mankind can see the glory of Creation.

If you have a science idea that you are passionate about, that is accurate and true, you will be remembered fondly by the world; but if your scientific hypothesis is wrong, then eventually your blunder will become apparent. Your folly will be made known, and your false reputation will pay dearly whether you are dead or alive. So, always make sure you consider the opposing perspectives and re-check and adjust your own thinking, when needed, so that you are *accurate*.

We do **not** have it all figured out! Science has barely scratched the surface of the heavens. And the heavens are always waiting for a new generation of scientists to discover all that we see and all that we do not see. Who knows what wonderful contributions to humanity *you* will come up with while trying to see what *is*.

The patriarchs of science are sprinkled throughout history, leaving a trail of discovery behind for us all to enjoy! Because of their discoveries we need not make the same mistakes that they had to endure in order to draw their objective conclusions. These people had a passion for their work, and while we can imagine that validation from peers and the general community was welcomed by them, based upon many of their own words, these patriarchs cared little for the spotlight if the truth was not in their announcement. Their passion was not to prove that their idea was right; their passion was to accurately explain what they observed as best as they could; and from that they gained their deserved recognition. If you want to be a "science-great", then being a science-great should *not* be your passion—seeking truth should be your passion!

We can see the stars, but the rest, at this point, is mostly guesswork because we have only just begun! *Science* is the discovery and finding of what *IS*.

Viva La Science!

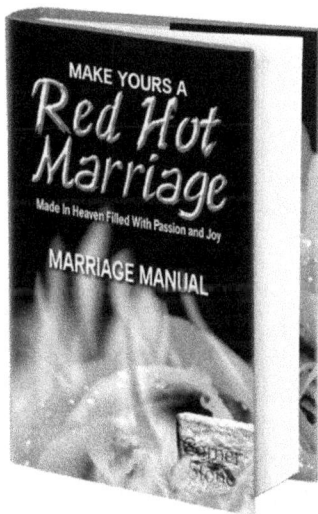

The Science of God
The First Four Days
Volume 1 - The First Four Days

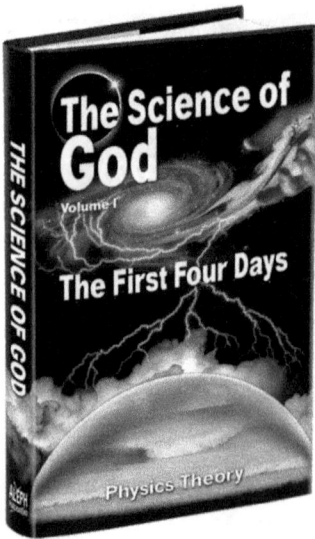

Is there a God? Did we evolve? Did everything start from a big bang? These questions have been plaguing our minds for many years. Only science-minded people and clergy seem to have the answers. But do they really have any true answers?

Is what we are told by science true? Is what we are told by the Church true? Or are there other better explanations for everything? Did we hitch a ride from Mars, or is that all fantasy science? Was everything Created in six twenty-four hour days, or did it all take billions of years to happen? Few people are willing to even fully consider these questions, and even fewer have any coherent answers. *The Science of God* challenges your current beliefs while asking tough questions of science and of the Church.

For years, Christian after Christian has attempted to argue for God and the Bible's Creation only to fail miserably. Why is this, why is it that Christians cannot seem to win this debate? Often Christians think they are winning the debate only to find themselves at a loss to answer the real questions, and then they get mocked for their poor answers.

Whether you are a scientist or an average Christian and want to discuss the Creation debate, *The Science of God* is a mandatory read for you. *The Science of God* takes you through the thought process to enable you to speak intelligibly about Creation, the cosmos, evolution, and astrophysics.

Search: The Science Of God Book
SayItBooks.com

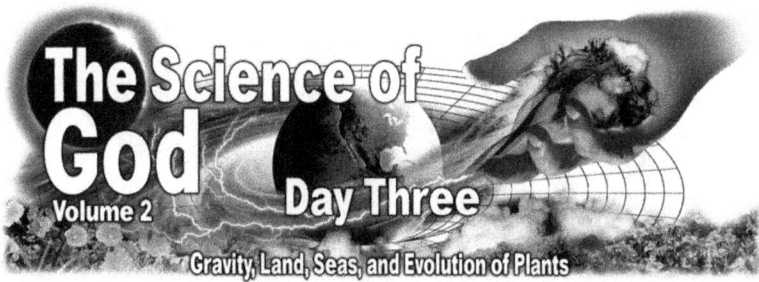

UNDERSTANDING THE CHURCH

Upon This Rock I Will Build My Church

Church in the Lurch - a House Built Upon Sand

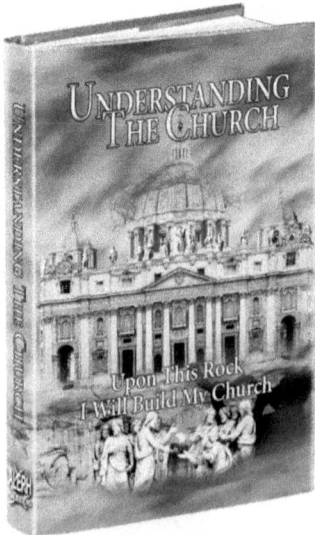

The Church is rapidly dying, and much of the clergy in recent times have been doing it more harm than good. People are fleeing from the Churches as they seek a religious perspective that fits a modern worldview. Should we revive this old Church and try to save it from its own demise? What exactly is "The Church", and who or which of the many religions is the official caretaker of it?

The Christian religions of the world have done their fair share of damage to themselves and to the world, but in the bigger picture, they have done more good than damage. Saving the Church is probably worth our collective efforts because the Churches are perhaps the most charitable group of organizations that existed throughout history and even up to today.

The main reason that the Churches are in the rough condition that they are today is due to a lack of understanding by clergy and congregation. We can overcome this dark era of the Church and revive it only through *Understanding The Church*.

Understanding The Church will help you in Bible study, or even to simply better understand the Church. But most importantly, *Understanding The Church – Upon This Rock I Will Build My Church* will help to revive this dying patient.

Search: Understanding The Church Book
SayItBooks.com

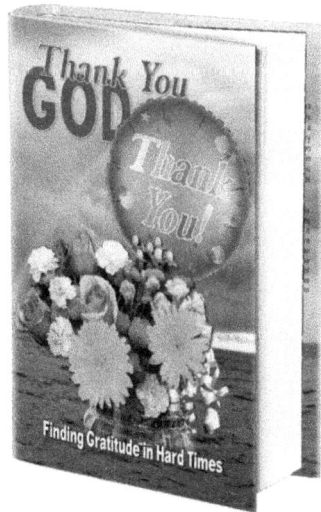

Notes

Notes

Notes

www.ingramcontent.com/pod-product-compliance
Lightning Source LLC
Chambersburg PA
CBHW071409090426
42737CB00011B/1407